Marine Conservation for the 21st Century

MARINE CONSERVATION FOR THE 21ST CENTURY

by Hillary Viders

BEST PUBLISHING COMPANY

Book Design & Layout by Peter Fine

Published-1995

ISBN: Paperback 0-941332-46-2, Hardback 0-941332-47-0
Library of Congress Catalog Card Number: 95-078019

Composed, printed and bound in the United States of America

Best Publishing Company
2355 North Steves Blvd.
P.O. Box 30100
Flagstaff, AZ 86003-0100 USA

TABLE OF CONTENTS

FIGURES

INTRODUCTION

During the course of our lifetime, most of us will spend time in or around the ocean, and in doing so, will glimpse the beauty and mysteries of some of our planet's most diverse ecosystems. With this privilege, however, comes responsibility to preserve and enhance this extraordinary environment, not only for our personal enjoyment, but more importantly, as a legacy to future generations.

Marine Conservation for the 21st Century is a vehicle to educate people in marine ecology, and to motivate people to become actively involved in today's environmental issues. This book was written because dedicated divers, boaters, fishers, scientists, educators, resource managers, and conservationists around the world have expressed their need and desire for it.

I would like to thank the individuals who reviewed the text and offered many helpful suggestions: Dr. Jack Pearce, Dr. Sylvia Earle, Dr. James Bohnsack, Dan Orr, Betty Orr, and Billy Causey. I am especially grateful for the help which was extended to me by the wonderful staff at the Center for Marine Conservation (CMC), including Roger MacManus., Rose Bierce, Kathy O'Hara, Bruce Ryan, and Linda Maraniss. Some technical information was reprinted, with permission from CMC, from academic materials by Mike Weber, Richard Tinney, and Susan Fowler. Illustrations were done by Keith Ibsen. I am also indebted to the many photographers who contributed photos to this book, particularly Stephen Frink, Mort and Alese Pechter, and Pete Nawrocky.

Most of all, I thank the two greatest sources of love, inspiration, and encouragement in my life—my husband, Richard, and my son, Jordan.

—Hillary Viders

NOTES

WHAT IS MARINE CONSERVATION AND WHY IS IT SO IMPORTANT?

"For most of history, man has had to fight nature to survive; in this century he is beginning to realize that, in order to survive, he must protect it."
—*Jacques Cousteau*

Marine conservation is one of the more important issues in the world today. A literal definition of "marine conservation" would be: "preventing the loss, decay, destruction, or injury of the oceans." But this is too simplistic; in reality, there are many complex factors which directly and indirectly contribute to the degradation of our aquatic world, which includes marine as well as fresh water environments.

Historically, water has always been considered a renewable or infinite resource, and therefore has been ignored, abused, and often poorly managed, with little regard for the consequences. Well into the twentieth century, the feeling persisted that there would always be plenty of wilderness left, and that man had been given the innate privilege to take whatever was there with impunity. But in modern times, we have discovered that water is finite, and that in many regions of the world, usable water has become a perilously scarce commodity. The tragedy is that this realization has not yet translated into sufficiently effective action.

The marine environment is becoming increasingly stressed by burgeoning population and industry. Our current world population of 5.4 billion represents a 200% growth rate in only the last forty years. Without intervention of famine, disease, or war, scientists estimate that the world population will double to 10.8 billion by 2045, and could almost triple to 14 billion by the end of the next century. The bottom line which all of us must face is, "How many footprints can the earth hold?"

Conservation seeks to harmonize the biological, social, and economic needs of our planet's inhabitants. One of the outstanding

–Photo: Mort & Alese Pechter

The oceans and waterways of the world are a primary area
for recreation.

problems for conservationists today, for example, is to reconcile the
different uses to which human beings aspire to put the oceans. For
example, how can we successfully exploit coral reefs for tourism
and recreation, while not despoiling them as a prime food source?

Many people find the topic of marine conservation intimidating
because its scope and complexity seem overwhelming. But the
majority of these problems are caused by human activities, and
individually and collectively, people can exert various degrees of
control and change.

Conservation necessitates the education and active involvement
of scientists, politicians, economists, and educators, as well as citizens. Skin and scuba divers, fishers, boaters, and other recreational
groups have a multi-dimensional investment in problems that affect
the marine environment. These user groups not only want a clean
and safe environment for living, but the enjoyment and safety of
water sports depends on the biological and aesthetic integrity of the
aquatic environment. For the large number of professionals who
earn their living in aquatic industries, the decline or impoverishment of marine and freshwater environments means financial
disaster.

Recreational users are often the first to notice problems in the
aquatic environment, and divers' observations and reports have
been invaluable to marine scientists and management agencies. It is

Divers have an intimate glimpse into some of the most fascinating and diverse ecosystems on this planet.

Photo: Mort & Alese Pechter –

Field trips provide an excellent way to get hands-on environmental education.

– Photo: Laddie Aikins/REEF

particularly important for the lay public, therefore, to understand environmental issues and to participate in the decision making processes. This book will present an overview of aquatic environments and principles of ecology as they apply to aquatic environments, a discussion of the problems in marine and freshwater ecosystems, and an examination of conservation initiatives currently underway by government, private sectors and grassroots groups. Most important, this book provides a plan for corrective and preventive action in which every person–whether a scientist, an aquatic sportsman, a legislator, or an ordinary citizen–can play a vital role.

WHAT IS A CONSERVATIONIST?

A conservationist is an individual who:

1. appreciates the basic elements and interactions in ecosystems;
2. is knowledgeable about the natural and human-induced impacts affecting ecosystems;
3. recognizes that how mankind lives and acts determines the fate of all our planet's living resources–both terrestrial and aquatic;

There are numerous field trips which teach people abut the marine environment and conservation. Pictured is a whale watching expedition.

Photo: Nina Young/courtesy of the Center for Marine Conservation–

–Photo: Mort & Alese Pechter

Marine education is important for people of all ages.
Pictured: The Discovery Center in Ft. Lauderdale, Florida

4. avoids and discourages behavior which causes negative effects to the environment;
5. organizes and participates in activities which have a positive environmental impact;
6. prefers long-term sustainability of resources, instead of short-term economic profits; and
7. has a strong sense of stewardship for protecting and preserving our global legacy for future generations.

PRIMARY AVENUES FOR PROMOTING MARINE CONSERVATION

There are three primary methods by which people can become involved in and promote marine conservation:

I. ENVIRONMENTAL EDUCATION

Learning about and recognizing the basic components and relationships in aquatic ecosystems, and the forces which can disrupt their natural balance. This knowledge can be acquired in college courses, lectures and symposia, and self-tutorial materials. If we are to act in the best interests of nature, then we have to understand as best we can how the planet works and how its creatures survive. Therefore, it is important that all of us, particularly educational leaders, obtain a basic knowledge of aquatic environments and problems in conservation. Primary avenues of environmental education are:

1. courses and lectures offered by colleges and universities;
2. local, regional, and national environmental organizations;
3. environmental programs within dive stores, dive clubs, local libraries, schools, and religious and community centers;
4. films, videos, books, magazines, and newsletters that discuss marine ecology and conservation issues. Some of the most informative publications are produced by oceanographic institutions and environmental organizations, such as *Oceanus* (published by Woods Hole Oceanographic Institution), *National Geographic, Marine Conservation* (published by the Center for Marine Conservation), *Sanctuary Currents* (published by NOAA), *Sea Technology, Underwater Naturalist* (published by the American Littoral Society), *Calypso Log* (published by The Cousteau Society), and *Living Oceans News* (published by National Audubon Society) (See complete list of recommended publications in Appendix 4.);
5. environmental research programs, projects and field trips. Many areas in the U.S. and abroad, now offer visitors opportunities to participate in a broad spectrum of marine science and conservation experiences–whale watching, reef restoration, fish and wildlife surveys, etc. (See chapter 18 on Ecotourism.);
6. undersea and marine science conferences and forums, such as the AAUS (American Academy of Underwater Sciences) are opportunities to meet and speak with marine experts; and
7. programs, exhibits, and materials in libraries, maritime museums, and aquariums.

II. MINIMIZING IN-WATER IMPACTS

Learning and practicing in-water skills and behavior that minimize environmental effects.

III. PARTICIPATING IN CONSERVATION ACTIVITIES

Personally participating in and encouraging others to participate in conservation activities, continuing environmental education, and conservation-minded citizen action.

THE MARINE ENVIRONMENT

"All is born of water, all is sustained by water."　　　*—Goethe*

The oceans and waterways of our world have been the most abundant source of life for over three billion years. Water is necessary for almost every project and process in nature and civilization. Water makes up two-thirds of the human body, and it is so vital to survival that we cannot live more than a few days without it.

Water is eminently suitable to support life because of its extraordinary physical and chemical properties. For one, it is an excellent building material–extremely flexible, but not very compressible. Water is a compound composed of two atoms of hydrogen attached to one atom of oxygen: H_2O. Because of its particular electrochemical nature, in which the molecules tend to break apart and carry electric charges, water is an ideal conductor for many of life's metabolic processes.

Water is also an excellent solvent: more different kinds of atoms, ions (charged atoms) and molecules can be dissolved in water than in any other naturally occurring material.

Not the least of water's important properties is its remarkable capacity to absorb heat–a greater capacity than that of any other material except ammonia. It takes one calorie of heat energy to raise one gram of water through one degree Celsius. A typical living cell contains almost eighty percent water; its ability to absorb heat without raising temperature protects all the delicate organic molecules that it contains.

Another fortunate property of water is that although it can freeze at 0°C and boil at 100°C, it is liquid at precisely the kinds of temperatures at which complex organic molecules, such as protein and DNA, need to survive and be chemically active.

*The "water planet,"
earth, as it appears from outer space.*

Photo: courtesy NASA–

In the late 1960s and 1970s, astronauts on the moon saw clearly what no one had observed so well or so dramatically before–that the earth is a "water planet" and that over seventy percent of the earth's surface is covered with water. There are approximately 325 trillion gallons of water on the planet earth. The oceans comprise ninety-seven percent of the liquid mantle, with most of the remainder locked in polar icecaps. The Pacific Ocean alone is twenty-five percent larger than all of the world's land surfaces combined. The oceans occupy over 300 million cubic miles. Only the tips of the continents show above the ocean. Beyond the edge of the present shoreline, and extending out to sea, are the true edges of the continents. The submerged fringe of each continent is the CONTINENTAL SHELF, which may range over hundreds of miles in width in some areas. At the outer edge of each shelf, the sea floor plunges precipitously down into the depths. The average depth of the oceans is over two miles, with the greatest known depth being just over seven miles in the Marianas Trench of the western Pacific.

Even the shape of the ocean bottom was all but unknown until recently. Concerted efforts at mapping the ocean have shown us that the topography of the ocean floor is even more spectacular than that of land. The mountain ranges and canyons of the oceans are higher, deeper, and larger than those on land. The seaward edge of the underwater continental shelf is scored in many places with submarine canyons.

Since the beginning of time, the marine environment has shaped the very character of our planet, and has profoundly influenced human civilization. The marine environment is a trove of energy resources, valuable minerals, and countless living organisms.

Oceans and related waterways provide food, livelihood, and habitats for a majority of the planet's life forms. They influence our weather patterns, constitute a major avenue for world trade and travel, and are vital to fundamental scientific and medical research. It would be impossible for the vast majority of earth's creatures, including humans, to thrive without healthy seas.

–Photo: Mort and Alese Pechter

The oceans of the world contain an incredible variety of creatures and activities. Pictured: A tiny sharpnose goby peers out from his shelter.

The aquatic environment contains fascinating creatures. Pictured: A squid up close.

Photo: Keith Ibsen –

The marine environment also provides some of our greatest recreational resources. Millions of people visit the sea every day. The marine environment is a haven for swimming, boating and sailing, skin and scuba diving, fishing, clamming and crabbing, surfing, camping, and photographing wildlife.

A Historical Perspective of the Oceans

"For all at last returns to the sea–to Oceanus, the ocean river, like the ever-flowing stream of time, the beginning and the end." **–Rachel Carson**

Scientists believe that the oceans formed over a period of millions of years as continents were pushed apart by great rifts in the earth's surface. Surprisingly, the water of the oceans and some of the life forms in it are older than the sea floor itself. As the earth cooled in the first years of its existence, water vapor in the atmosphere condensed and rained down upon the earth. Over millions of years, more and more water vapor was added to the atmosphere as volcanoes erupted and sent great plumes of gases, including steam, into the air. Slowly, water accumulated on the earth's surface. At the same time, this surface was constantly changing. Volcanic activity pushed rock from the center of the earth to the surface where it formed vast plates of new crust. Older crust was degraded at places where a crusted new plate was forced over it.

Over millennia, life proliferated in the oceans. Amazingly, sharks, species that developed hundreds of millions of years ago, still exist today. In fact, the oceans teem with "living fossils"–plants and animals which evolved eons ago and remain suitably adapted to the present environment. So agreeable are the oceans to life that all ocean waters, no matter their depth, darkness, or temperature, support some kinds of plants or animals, sometimes in astonishing variety and numbers. Animals including fishes, seabirds, marine mammals, mollusks (clams and snails), crustaceans (lobsters, shrimps, crabs, and barnacles), and echinoderms (sea stars and sea urchins) exist in a profusion of forms in all the parts of the oceans. Interestingly, insects are almost completely absent from the oceans.

Humans began harvesting the plant and animal bounty of the oceans before the beginnings of recorded time, using marine resources for food, clothing, and art objects. As cultures became

−Photo: Paul Humann/REEF

Healthy living oceans are necessary for human life.

more sophisticated, they also became more daring in their exploitation and exploration of the sea. Around 2,000 BC, Egyptian voyagers
were traveling as far as Arabia and Somalia. A thousand years later
the Phoenicians were exploring the Mediterranean, the Red Sea, the
Gulf of Aden, the Persian Gulf, and the Arabian Sea, venturing at
least as far as India. They also sailed through the Straits of Gibraltar,
discovering the Canary Islands in the Atlantic.

For the most part, ancient civilizations had little scientific
knowledge of the oceans, often mixing fact with myth and folklore.
But by 325 BC, Aristotle was operating a marine biological site on
the island of Lesbos in the Mediterranean Sea. Aristotle, famous
philosopher and tutor of Alexander the Great, described the anato-
my and behavior of 180 species of marine animals, and some of his
classifications are still considered valid. Around 100 BC,
Poseidonius, a Greek geographer, measured 6000 foot depths of the
Mediterranean near Sardinia and studied the tides at Cadiz, corre-
lating them with the phases of the moon.

In the meantime, the people of the Pacific independently developed their own navigation techniques, which included star charts
and knowledge of currents and bird migration. (See Figure 1)

With the discovery of the New World by Europeans, marine science and marine industry took a great leap forward. The "Age of Navigation" dawned and the oceans became rivers of commerce. The wealth of the colonies in the New World and the Far East was exchanged for manufactured goods from Europe.

Together with the development of trade came increased knowledge. Modern ocean science can be said to have started with Portuguese and Spanish navigators in the fifteenth century. As they ventured forth and farther, they discovered many of the ocean's major currents. They also needed an accurate means of determining their location. So, over many years, the science of celestial navigation developed.

In the United States, ocean science began with Benjamin Franklin's charting of the Atlantic Ocean's Gulf Stream. Knowledge

– Illustration: K. Ibsen

Figure 1 – A Polynesian Star Map

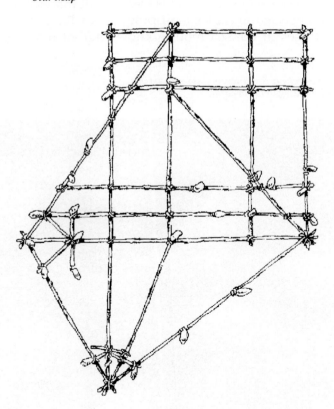

of ocean currents advanced again in 1855 when Matthew Fontaine Maury, a U.S. Navy navigator and Superintendent of the Navy's Depot of Charts and Instruments, published his *Physical Geography of the Sea*, the first major oceanography text.

About the same time, marine biologists were unlocking some of the secrets of life in the sea. One of the earliest voyages of biological discovery was the 1831-1836 expedition of the H.M.S. Beagle on which Charles Darwin garnered data that led him to his theories of natural selection and evolution. Darwin also developed a theory of the origin of coral atolls, a theory still widely accepted.

The greatest oceanographic expedition which had been undertaken up until that time was the 1872-1876 voyage of the H.M.S. Challenger. On a winding, seventy thousand nautical mile voyage around the world, the Challenger expedition gathered information on the physics, chemistry, geology, and biology of the oceans. The results of the voyage filled fifty large books and took twenty-three years to publish.

As in the past, today's use of the oceans depends on an understanding of these vast expanses of water and all that occurs in them. Fisheries, transoceanic cables, shipping, oil drilling, and all other commercial uses of the sea require knowledge of many aspects of

The aquatic environment contains complex ecosystems.
Pictured: A goby hiding in brain coral.

–Photo: Mort & Alese Pechter

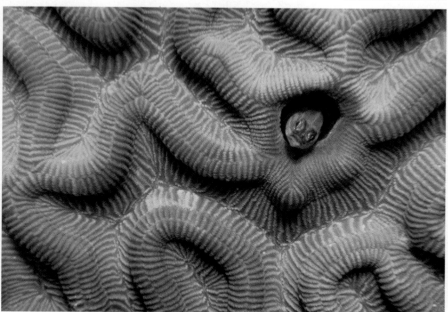

–Photo: Mort & Alese Pechter

The aquatic environment contains fascinating creatures. Pictured: A manatee

marine science. The same is true of military uses of the sea and even of some recreational uses such as scuba diving and sailing.

With satellites, remotely-piloted submarines, DNA probes, and sophisticated electronics, ocean science has entered the high-technology era. Every year, science continues to uncover hundreds of new life forms in the sea. Progress in sampling techniques, biochemistry, and microscopy has expanded our ability to find and identify life in the sea. In 1960, by using *Trieste*, a deep diving bathyscope, two scientists were amazed to discover fish living at the bottom of the Marianas Trench, a near freezing abyss devoid of light, where ambient pressure is about 16,000 pounds per square inch.

In 1966, the world held its breath as pilots from the Woods Hole Oceanographic Institution (WHOI) submersible *Alvin* recovered a hydrogen bomb which had fallen into the Mediterranean after two foreign military planes collided. In the following decades, WHOI has used deep sea submersibles to explore such diverse topics as geological activity deep within the earth, plant and animal populations and their interaction in the oceans, coastal erosion, ocean circulation, pollution control, and global climate changes. Wood's Hole Oceanographers are credited with having uncovered vast underwater mountain ranges, new life forms in the Galapagos Islands, hot water springs miles beneath the ocean surface, and with the development of sonar that transmits clear images from an inky black ocean depth of 10,000 feet. In addition to the vessel *Asterias*, WHOI's research fleet now includes the 279 foot *Knorr*, 210 foot *Atlantis II*, 177 foot *Oceanus*, 65 foot *Eagle Mar*, the three-person submersible *Alvin* which is capable of diving to 13,000 feet, and ROVs such as *Argo* and *Medea/Jason*.

Photo: Keith Ibsen—

Divers enjoy an intimate relationship with the sea and its creatures. Pictured: A diver swimming with a whale shark.

Dr. Sylvia Earle in a submersible exploring the deepest depths of the ocean.

In the 1990s, Japan emerged as a global leader in deep sea exploration. Japan, a country which is literally dominated by the marine environment, is eager to probe the abyss to understand the movement of its tectonic plates, and hopes that the information will help scientists predict the earthquakes, volcanoes, and tsunamis which have historically plagued the country. The deepest manned submarine in the world, the *Shinkai*, recently set a depth record in the Atlantic, carrying three people down nearly four miles. In 1994, the Japanese robot, *Kaiko*, set a depth record of seven miles.

Dr. Sylvia Earle, co-founder of Deep Ocean Engineering in San Leandro, California, and the world record holder for a solo deep ocean walk (1250 feet), helped design the *Deep Rover* submersibles. Dr. Earle is currently involved in "Ocean Everest," a project which will explore the ocean's deepest recesses by means of innovative hi-tech submersibles, *Deep Flight*.

In another current project, ships equipped with drills, manned submersibles, and unmanned robots are probing deep volcanic vents two miles down in the Atlantic Ocean. The vents, some dating back 50,000 years, spurt plumes of black sulfurous smoke and 685°F water. The largest of the huge volcanic mounds currently being explored lies 200 miles east of Miami, on the Mid-Atlantic Ridge. The research is expected to reveal information about how deep vents support a myriad of marine life, and how they produce rich metallic ores (including gold).

New discoveries, however, are by no means confined to the ocean floor. Since 1988, researchers at the Monterey Bay Aquarium Institute in California have been using manned submersibles and robots to explore the middle depths, where they have been uncovering new types of animal life at a rate of about twelve species a year. The middle regions of the ocean are enormous and vastly unexplored. By volume, land is estimated to make up about 0.5% of the earth's habitable space. The sea comprises the other 99.5%. Dr. Bruce Robinson, the chief scientist of the project, contends that the middle ocean range contains the most dominant life forms on our planet in terms of biomass, numbers of individuals, and geographical extent.1

THE PHYSICAL ENVIRONMENT
OF THE OCEAN

"Nature's craft and ingenuity are nowhere more evident than in the delicately spun web of life held within the oceans." **—The International Oceanographic Foundation**

The physical environment and topography of the ocean influence the distribution of its living inhabitants. Both the continental shelf and the seafloor sloping to the deep ocean plain may be rocky, gravely, sandy, or muddy. Each type of surface presents its own challenges to organisms living there. Many of the same challenges are found in the water itself, for the physical environment of the open ocean is by no means uniform.

LIGHT

Differences in light, water temperature and movement, salinity, and pressure create barriers between the waters at different depths. Scientists use vertical and horizontal divisions of the ocean in discussing various ocean environments. (See Figure 2.) For instance, the uppermost zone of the ocean, into which sufficient light penetrates to allow plant growth, is called the euphotic zone (eu=good, phot=light). In turbid coastal waters this zone may be 90 feet (270 meters) or less in depth; in clearer mid-oceanic waters, the euphotic zone may extend down to 600 feet (1800 meters). The amount of light penetrating the water gradually diminishes with depth until none at all is able to get through. This marks the beginning of the aphotic zone (a=without). The aphotic zone makes up over ninety-five percent of the oceans.

SOUND

Contrary to Jacques Cousteau's poetic description of the sea as a "silent world," the oceans are bristling with noise. Sound travels so

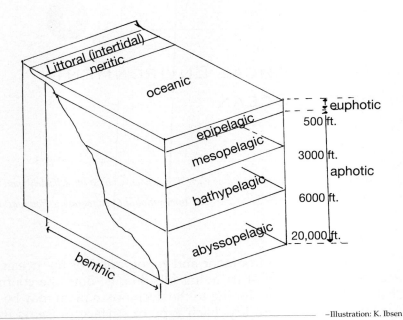

−Illustration: K. Ibsen

Figure 2– Principle vertical and horizontal divisions of the oceans

well in the water that low-frequency sounds can be heard for distances of several thousand miles. Underwater acoustics, therefore,
is an important field of oceanographic research. At Woods Hole
Oceanographic Institution, researchers have identified and recorded
energetic, indeed often bombastic, sounds produced by over forty
species of marine mammals alone-clicks, squeals, moans, and
grunts which are used for inter-species communication. The exact
meaning of these sounds and their patterns is still largely unknown.
At the Marine Physical Laboratory at Scripps Oceanographic
Institution, researchers have been using sonar research to gain a
better understanding of ocean currents and the physical properties
of ocean water masses. They recently developed a powerful new
technique, "acoustic tomography," in which selected areas of the
ocean are implanted with buoys containing high-tech acoustic
instruments (transducers). These transducers received and transmit
sound signals as deep as three thousand feet underwater.

PRESSURE

Depth increases the pressure exerted by water on organisms.
Many marine animals, from earthworms to octopuses, rely almost
entirely upon hydrostatic pressure to create their body forms. The

air pressure at sea level is 14.7 pounds per square inch (psi) or one atmosphere. This is the weight of a column of air, one inch square, reaching from the ground to a height of about 100,000 feet (300,000 meters). Because sea water per unit is about 800 times heavier than air, the pressure per square inch increases one atmosphere for each 33 feet of additional depth in sea water. Thus, the water pressure at 6,000 feet is nearly 300 times greater than the pressure we feel when snorkeling at the surface. Despite these great pressures, ocean depths are by no means without life. The adaptations of animals in these zones to the total lack of light and to the tremendous water pressures are among the most remarkable in the animal kingdom.

WATER TEMPERATURE AND MOVEMENT

The sun affects ocean environments in other ways. Because of the curvature of the earth and the atmosphere, polar regions receive less sunlight than do equatorial regions. Both the air and the surface water of tropical regions are, therefore, warmer than those of polar or temperate regions. To maintain the broad atmospheric heat balance of the earth, the warm air of the tropics tends to move toward the colder polar regions. (See Figure 3) As it rises and moves toward the poles, this warm tropical air cools and becomes more dense, cre-

Figure 3–Direction of Air Currents and Rotation of Earth

Illustration: K. Ibsen–

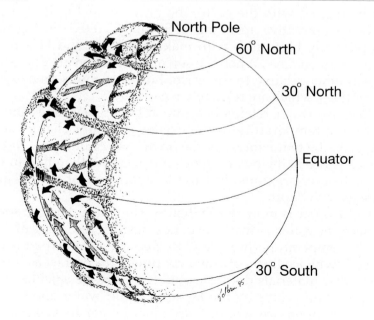

ating a high pressure area in polar regions. The cooler and heavier polar air moves toward the low pressure in the tropics, creating a relatively steady wind from the north in the northern hemisphere and from the south in the southern hemisphere.

The movement of these air masses is complicated by the rotation of the earth. Due to an effect of the earth's rotation that deflects all moving bodies to the right (known as the Coriolis force), a warm tropical air mass in the northern hemisphere that is moving toward the north pole is deflected to the right and eventually forms a gyre (a circular course or motion) turning clockwise. Some of the cooled tropical air flows beyond the gyre, and becomes part of a second, temperate gyre.

The movement of these air masses creates wind forces, such as trade winds, which in turn drive ocean currents. The easterly winds that blow near the equator, for instance, drive the great equatorial currents of the ocean. As these currents move westward, they are deflected to the right by the continents. These major currents and the many counter-currents and eddies circulate the water in the oceans, the mineral elements in the water, and much of the plant life upon which other life forms in the oceans depend.

Another factor in movement of ocean water are tides. It takes the earth on its axis, twenty-four hours and fifty minutes to complete a rotation, during which time the ocean undergoes two high tides and two low tides. Tides vary in strength, depending on the position of the sun and moon. When the sun and the moon are nearly aligned with the earth, the gravitational pull of these two forces is cumulative, resulting in SPRING TIDES. Spring tides are exceptionally high tides with maximum rises and falls occurring every fourteen days. When the sun and the moon are at right angles to each other relative to the earth, however, the oceans experience NEAP tides, which are relatively weak tides.

Because water retains heat much better than does the atmosphere, the temperature of the oceans does not change abruptly. However, the temperature of the oceans is not uniform. Not only is surface water in the polar regions much colder than that in temperate or tropical regions, but the temperature of the oceans also changes with depth.

Cold water is more dense than warm water. Another fortuitous property of water is that it expands, just on the point of freezing. As the temperature drops to O degrees C, water expands by 10%, which is why ice floats. If water continued to contract as it froze, as most other materials do, ice would sink, and be shielded from the Sun's warming rays by the layer of liquid water above it. If this were to happen, ice would take much longer to melt, and many more of the world's waterways would be permanently frozen. Similarly, more saline water is more dense than less saline water. As a result, the ocean is layered with water masses of different salinity and temperature. These water masses can move in different directions and at different speeds. But more important, the boundaries between such layers can form effective barriers to migration between the layers by marine organisms. Some organisms cannot tolerate the temperature or salinity differences that would be encountered in moving from one water mass to another. All of these variables create a broad variety of environments for life in the ocean.

There is a constant recycling of water powered by the Sun which is referred to as the HYDROLOGICAL CYCLE. In the hydrological cycle, approximately 300,000 cubic kilometers of surface ocean water evaporate every year. About two thirds of this water vapor falls back into the sea as rain or snow; the rest falls onto land. The water that reaches land travels through rocks into aquifers or flows as surface rivers into the sea, propelled by the Earth's gravity. Thus, a considerable part of the rainfall that supports land ecosystems and most human food production comes from the oceans.

PHYSICAL DIVISIONS OF THE OCEAN ALONG THE U.S. COASTLINE

Based on the variables mentioned above, biologists have divided the coastline of the continental United States into seven physical regions, or oceanographic provinces:

1. Acadian province–extends from Newfoundland and Nova Scotia to Cape Cod;
2. Virginian province–extends from Cape Cod to Cape Hatteras, NC;
3. Carolinian province–extends from Cape Hatteras to Cape Kennedy, FL;
4. Louisianian province–extends along the Gulf of Mexico;
5. West Indian province–extends from below Cape Kennedy on the eastern coast of Florida to Cedar Key on the western coast of Florida, and southward from Port Aransas, Texas, to the southern portion of the Gulf, the Yucatan Peninsula, Central America and the Caribbean Islands;
6. Californian province–extends from Mexican waters north to Cape Mendocino, CA; and
7. Columbian province–extends from Cape Mendocino northward to Vancouver Island, British Columbia.

NOTES

PRINCIPLES OF ECOLOGY

"The study of ecology gives us more than practical knowledge. It also shows us the earth's fantastic beauty and the incredible variety of life."
— **Eugene Odum**

To appreciate the aquatic environment in its broadest sense, it is necessary to understand its smaller components and dynamics, or ecology. Ecology (eco=household, ology=study of) is defined as the scientific study of interactions of living organisms in their environment. Because biologists now calculate that there may be as many as fifty million different species of living creatures on earth, and because the physical environment is more intricate than we can ever measure, ecology has emerged as an infinitely complex science.

In the broadest sense, ecological issues were debated as far back as the Middle Ages, when people complained about too much smoke spoiling the environment in large cities. Ecology, however, as a formal science, is relatively young. The word "ecology" was only coined in 1873. Although the laws of ecology are still being developed, some of its principles have already won wide acceptance. The following are some key principles of ecology which enhance our understanding of the marine environment:

I. A COMMUNITY AND ITS ENVIRONMENT—THE LIVING AND THE NON-LIVING—CONSTITUTE AN ECOLOGICAL SYSTEM, OR ECOSYSTEM.

A SYSTEM is defined as a "regular interacting or interdependent group of items forming a unified whole." A living system is complex. It is composed of component subsystems that process information, matter, and/or energy. We study ecology at two levels, the ecology of individual SPECIES and that of ECOSYSTEMS.

The Swedish biologist Carl Linnaeus coined the term SPECIES in the eighteenth century, as part of his modern sys-

–Photo: Mort & Alese Pechter

The principles of ecology help us understand and protect marine ecosystems.

tem of classifying living things. Each "kind" of creature is a species. Linnaeus also grouped similar species into larger groups, called GENERA (singular genus). Genera are grouped into FAMILIES, families into ORDERS, orders into CLASSES, classes into PHYLA (singular, phylum) and phyla into KINGDOMS. Each species was given two names, the name of its genus plus its own specific name.

An ecosystem is a unit that includes all of the organisms in an area interacting with their physical environment. These interactions involve a network of biological, chemical, and physical changes. An ecosystem can be as large as the earth itself, or as small as a tide pool. Ecosystems exist on many kinds of land, and in oceans, lakes, rivers and coastal wetlands. Ecosystems are found wherever soil, air, and water support communities. Every natural community draws materials from its surroundings and transfers materials to it. Raw materials and decay products are exchanged continuously.

The factors operating in an ecosystem are divided into BIOTIC (living) and ABIOTIC (non-living) elements. Biotic factors include the plant and animal life of an ecosystem.

Abiotic factors include minerals, water temperature, sunlight and shade, average precipitation, nature of soil (for terrestrial ecosystems), altitude, latitude, salinity, dissolved oxygen, wave action, currents, tides, bottom type, amount of suspended particulate matter (for aquatic ecosystems), and seasonal changes. An ecosystem's LIMITING FACTOR is the element which controls the growth of that ecosystem (i.e., the limiting factor in a rain forest is availability of light).

The inhabitants of all ecosystems, whether aquatic or terrestrial, are classified into three categories:

1. PRODUCERS (sometimes called autotrophs, or self-feeders (auto=self; trophs=feeders);
2. CONSUMERS; and
3. DECOMPOSERS.

All green plants are producers, because they produce their own food. Animals, including fish and humans, feed on plants or on other animals and are therefore classified as consumers. Organisms that cause decay–bacteria and fungi–are decomposers.

Besides requiring energy, animals and plants require

Populations are the basic biotic factor in an ecosystem.

Photo: Hillary Viders–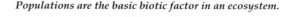

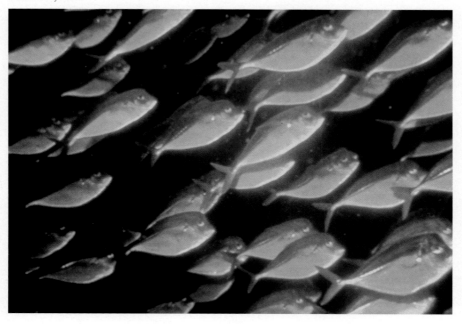

shelter, other nutrients, and reproductive mates. The manner in which these needs are met by various species of plants and animals in an ecosystem expresses some of the dynamics of that ecosystem (see Chapter 6, Functions in Ecosystems).

II. POPULATIONS ARE THE BASIC ELEMENTS OF THE BIOTIC FACTORS IN AN ECOSYSTEM.

A POPULATION is a group of individuals of the same species that live in a given area and interact with each other. A group of creatures who may or may not be of the same species, and simply happen to be in the same place at the same time, though not necessarily interacting, is an ASSEMBLAGE. The area where an organism or group of organisms live is their HABITAT.

It is only in recent decades that biologists have come to realize just how complicated populations really are: how many different forms a population may take, how many factors determine population size, and how vulnerable populations may be.

The size of a population within an ecosystem fluctuates depending upon a variety of factors. Extrinsic factors such as water temperature, predation by other species, and food sup-

A community may be made up of different species which inhabit a similar area. Pictured: A trumpet fish and a grouper.

–Photo: Keith Ibsen

ply, are largely beyond the control of the population. The intrinsic factors of competition for food, stress caused by crowding, and the presence or absence of reproductive mates are created within the population itself.

Like other aspects of ecosystems, populations are always adjusting to internal and external changes. Successful self-maintenance and self-regulation lead to an equilibrium or HOMEOSTASIS (homo=same, stasis-posture) in which the flow of energy and nutrients into and out of the ecosystem is balanced.

Regardless of size, communities are vulnerable in that when one inhabitant or relationship is removed, many others that are dependent upon it may disappear as well. When a prey-predator relationship changes, for example, the results may be far-reaching. It is for this reason that complex ecosys-

–Photo: Stephen Frink

Habitat is an important component of an ecosystem. Pictured: A coral reef is a habitat for numerous organisms.

An ecosystem is more than the sum of its parts. To understand a coral reef community, we have to study much more than the corals.

tems, such as coral reefs, are so fragile. Such an ecosystem's cross-links and inter-dependencies are so manifold that hundreds of species may disappear if only a few are removed.

Ecologists often use the term CARRYING CAPACITY to refer to the maximum population that can be supported indefinitely in a given habitat. Modern economists have also applied this term to human society. "Human carrying capacity," therefore, is used to describe the maximum rate of resource consumption and waste discharge that can be sustained indefinitely in a defined planning region without impairing ecological productivity and integrity.

III. CLOSELY RELATED TO THE LIFE PATTERNS PRINCIPLES IS THE PRINCIPLE OF BIOTIC COMMUNITIES.

A COMMUNITY is all collective populations living in a given physical habitat; the populations making up the community have a definite organization of feeding relationships and of transfer and use of energy. Often communities are classified according to dominant species as in the kelp-urchin-otter community. Communities may also be classified by the physical characteristics of the habitat, as, for example, the sandy beach community. While populations of different species may occupy the same habitat, they have different ecological niches. An organism's NICHE includes not only where it lives within a habitat, but also what it does and how it is influenced by other species. Over time, species with similar niches specialize and differentiate their needs so that competition for food, shelter, and space is reduced.

IV. IN NATURE, AN ORDERLY, PREDICTABLE SEQUENCE OF DEVELOPMENT, REFERRED TO AS ECOLOGICAL SUCCESSION, TAKES PLACE.

ECOLOGICAL SUCCESSION is a process whereby one form of life (plant or animal) replaces another, due to changes in the habitat. In simple terms, as time goes on, changes may take place either through natural or human-induced activities. In natural systems, succession is an ongoing process which is continuously changing, due to the conditions present at that specific moment in time.

Ecology teaches us that matter cannot be created nor destroyed. We can recycle, burn, bury, or dump our wastes, but they do not leave the planet. Pictured: A landfill operation.

Photo: The Center for Marine Conservation –

V. NATURAL SELECTION LEADS TO ADAPTATION.

In the mid-nineteenth century, Charles Darwin observed that when living things reproduce, the sibling offspring vary. He also noted that in a world in which many organisms are reproducing, there is bound to be competition for resources. Because the individual organisms vary, some are bound to be better able to survive in particular circumstances than others. Darwin referred to the process by which the fittest are singled out for survival as NATURAL SELECTION.

In natural selection, organisms that survive are the fittest in the sense that they are best adapted to the environment (but they are not necessarily the healthiest). The organisms that prevail, those that are selected for survival, are those that grow and reproduce most abundantly. These are the organisms that scavenge energy from their environment most efficiently and use it to repair and augment their own bodies and to produce facsimiles of themselves.

If natural selection, therefore, is the process through which living things become adapted to their environment, both physically and by their behavior, ecology is the study of the end results of that adaptation.

VI. THE COMBINATION OF ALL THE EARTH'S ECOSYSTEMS CONSTITUTES THE BIOSPHERE.

BIOSPHERE is the widely used term which refers to all the Earth's ecosystems (systems which encompass life on land, water, and air) functioning together on a global scale. The biosphere merges invisibly with the LITHOSPHERE (rocks, sediments, mantle, and the core of the earth), the HYDROS-PHERE (SURFACE AND GROUND WATER), AND THE ATMOSPHERE. Although the exact mechanism of these gigantic meshing gears is not completely understood, scientists believe that we can never affect just one ecosystem; whatever we do to one element in the natural world will have an effect on the other elements, including humans. Therefore, our conservation efforts must be holistic, that is, they should focus on preserving the biosphere and ecosystems in their entirety, rather than individual species.

VII. IN ECOLOGY, "THE WHOLE IS MORE THAN THE SUM OF ITS PARTS."

A forest, for example is more than a conglomerate of trees. Likewise, to understand the complexity of a coral reef, one must study much more than just the coral.

VIII. The First Law of Thermodynamics, the Conservation Law, states that the amount of energy and matter in the universe is constant.

Energy and matter cannot be created or destroyed; they can only be transformed from one state to another. Broad components of matter which undergo constant cycling are: carbon (photosynthesis and respiration), water (condensation, precipitation, evaporation), nitrogen, phosphorous, and sulfur.

When human interactions in the natural world generate wastes and pollutants, those materials are re-absorbed back into the earth. Our wastes can be buried in the ground, incinerated into the air, dumped in the water, can be recycled, or can remain as unusable waste, *but they can never disappear from the earth.*)

IX. Energy undergoes entropy.

ENTROPY, referred to in the Second Law of Thermodynamics, means that every time energy is transformed from one state to another (such as when fuel is burned), some of its ability to do work is lost. Entropy occurs when stored energy becomes cooler, less concentrated and/or less efficient when it is applied to work. One of the goals of conservation is to seek ways to minimize entropy, such as the use of the sun, a primary and natural energy source for light, instead of reliance on synthetic energy sources, such as incandescent light bulbs, which waste ninety-five percent of their energy input.

X. The earth is a "closed system" in which matter and energy are constantly recycled.

A open system is one which can exchange energy, information, and waste with its environment. The more open a system is, the more renewing it can be, thus it suffers less from entropy. This is because an open system is able to import sufficient amounts of energy from its environment to replenish what it loses when transforming its own energy, and it is able to expel waste products that result from this transformation back into the environment. The more closed a system, the less renewing it can be because it can neither import sufficient quantities of energy to replace its depleted sources, nor dispose of its waste. The more closed a system, the more entropy it suffers.

The earth is a closed system. This is why we need to examine and rethink our actions, not only in the water, but on land. The way we live day-to-day has an enormous impact on the entire biosphere and all its inhabitants.

FUNCTIONS IN ECOSYSTEMS

" The themes of life itself are continuity and change. "
— *George Reiger*

Although there are vast geographical, biological, and physical differences in ecosystems throughout the world, they share common denominators, i.e., basic functions which are mandatory for continuation. Activities in an aquatic ecosystem include the hunt for food and shelter, survival strategies, competition, predator/prey relationships, mechanisms of attack and defense, symbiotic and parasitic relationships, and reproduction and recruitment of species.

PRIMARY PRODUCTIVITY

Organisms, plant or animal, need energy for growth, reproduction, and other physiological processes. The source of almost all energy used by life on earth is the sun. In the oceans, PHYTO-PLANKTON (phyto=plant; plankton=floating) is the primary direct user of the energy of sunlight. Larger plants, such as blue-green algae, seagrasses, and seaweeds, also capture sunlight. These larger plants play a significant role in maintaining life in relatively small nearshore areas of the oceans, such as in seagrass and kelp beds.

All of these plants use the energy in sunlight to bind water and carbon dioxide into stored carbohydrates such as sugars, starches, and oils. In this process, called PHOTOSYNTHESIS (photo=light, synthesis=manufacture), plants convert the energy of sunlight into a form that is usable by other organisms. The rate at which plants fix or change the energy of the sun and make it more generally usable is known as PRIMARY PRODUCTIVITY, which determines to a very great extent the amount of animal life that can live in an area.

In addition to sunlight, plants need certain essential nutrients to grow. While salinity of the ocean is fairly constant at about 35 parts of salt per thousand parts of sea water, the availability of the nutrient salts upon which plant growth depends may vary. In many nearshore areas, for instance, necessary nutrient salts may be at very low concentrations for much of the year. During the rainy season, however, runoff from the land may increase significantly the availability of nutrients. Where there is intensive use of nitrogen and phosphorus-based fertilizers, large amounts of nitrates and phosphates may be washed into the sea, creating unusual blooms of phytoplankton. These blooms may rob other marine plants and animals of the sunlight and oxygen they require, and thereby impoverish the normal variety of life in that area. Some of the phytoplankton and detritus from other organisms sink to the ocean bottom, providing nutrients for bacteria and other benthic organisms.

Illustration: Keith Ibsen–

Figure 4-Typical Marine Food Chain

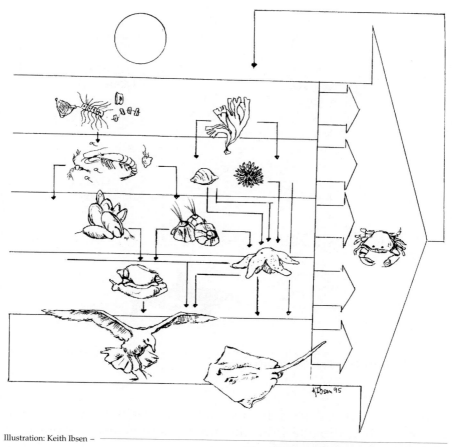

Illustration: Keith Ibsen –

Figure 5-Simplified Nearshore Food Chain

Additionally, some of the remaining nitrates, phosphates, and other nutrients may be brought to the surface by ocean currents. When this occurs, these nutrients become available once again to the phytoplankton in surface waters and the whole cycle begins again. Life in the ocean also requires oxygen to use the energy stored in carbohydrates. Oxygen dissolves into water easily, especially where the surface is stirred by the wind or waves. During photosynthesis, plants release oxygen into the water. In areas of prolific plant growth, the water is usually saturated with oxygen. Because cold water can hold more oxygen than warm water, deep water and polar water are often rich in oxygen.

Plants in the oceans also need carbon dioxide, which is found both in solid and gaseous forms, for photosynthesis. When plants and animals release the energy in carbohydrates through respira-

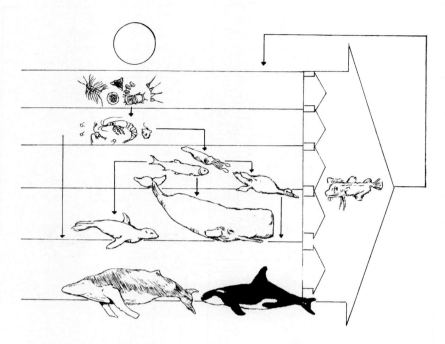

–Illustration: Keith Ibsen

Figure 6-Simplified Offshore Food Chain

tion, they produce carbon dioxide that dissolves in the waters of the ocean. Carbon dioxide from the atmosphere is also dissolved in water.

Although photosynthesis by green plants is responsible for the greatest production of the complex energy-laden compounds on which life depends, such compounds may be generated in other ways. Photosynthetic bacteria use hydrogen sulfide or organic compounds to create energy. Other bacteria do not even require light but create energy-laden compounds by the chemical oxidation of simple inorganic compounds, such as ammonia or sulfide. In this process, called CHEMOSYNTHESIS, bacteria that can live deep within sediments make mineral compounds such as nitrate available to other organisms. Without these bacteria, the minerals and the energy trapped in sediments would be lost to living creatures that depend upon them.

Because plants convert the energy of the sun into a form which can be used by other organisms, they are called PRIMARY PRODUCERS. These producers are preyed upon by a variety of primary herbivorous consumers (herb=grass, vor=eat), including ZOO-

PLANKTON (zoo=animal). Like all other animals in the oceans, zooplankton are not able to capture the energy of the sunlight directly and must depend upon plants to do this for them. The zoo-plankton may include small, shrimp like creatures and the larval forms of fishes and other marine animals. The herbivorous primary consumers are then preyed upon by secondary carnivorous consumers (carn=flesh), such as most fishes. Some of the secondary carnivorous consumers may be fed upon by tertiary consumers, such as wading birds. Finally, decomposers, such as bacteria and fungi, feed on detritus from these animals, completing the FOOD CHAIN. (See Figure 4)

Generally, each consumer species preys upon several other species. Likewise, an individual prey species is quite often preyed upon by several consumer species. These interrelationships are known as a FOOD WEB and illustrate just one of the complex rela-tionships among plants and animals in an ecosystem. (See Figures 5 and 6)

The farther removed a species is from primary production, the greater is its TROPHIC LEVEL (trophic=food). Only about ten per-cent of the energy at one trophic level can be captured by organisms of the next higher trophic level. Much of the energy captured by plants at the first trophic level is used by the plants themselves for

Marine creatures employ a variety of feeding strategies.
Pictured: A parrot fish uses its sharp beak to break off coral.

Photo: Mort & Alese Pechter–

respiration, the process by which plants and animals release the energy stored in carbohydrates. Only the energy left over after respiration can be passed on to the zooplankton at the second trophic level, and the process continues up the hierarchy. The way in which organisms obtain and use energy can reveal much about a marine ecosystem. Indeed, the cycling of energy among organisms is one of the principal issues an ecologist must address in any ecosystem.

THE SEARCH FOR FOOD

Life for many animals in the oceans requires constant search for food. As in other marine communities, coral reef organisms have adopted a dizzying array of feeding strategies. A variety of fishes, worms, sea stars, and other animals graze upon encrusting and fixed plant growth. Some groups of animals are fixed to the bottom and feed upon plant and animal material suspended in the water. Other members of the coral reef community, such as sponges and oysters, extract food from the water by filtering it through their bodies. Some creatures, such as crinoids, trap food with their tentacles. Reef fishes have many different feeding strategies. Parrotfish

The search for food is an essential activity in marine ecosystems. Pictured: A crinoid opening its tentacles to trap its prey.

–Photo: Hillary Viders

Photo: Stephen Frink–

*Prey/predator relationships are part of marine ecosystems.
Pictured: A green moray eel eating its prey.*

use their powerful teeth to rasp off encrusting algae and coral polyps. They then defecate the pulverized skeletons of the coral as sand. Butterfly fish have long beaks they use to pry into crevices, whereas sharks, barracuda, and jacks rely on their speed to capture prey. Groupers and sea bass, on the other hand, hide in the shadows; when prey wanders by, these fish thrust themselves from the shadows and swallow the prey whole.

COMPETITION

Competition is a characteristic of all communities. Plants compete for light and space, animals compete for space, food, and water.

Competition is greatest among living things that have similar needs and in areas where different types of communities overlap (i.e., wetlands). Animals and plants trying to establish a foothold in such an overlap must cope with difficulties nonexistent in a stable community. Sometimes competition is modified by behavioral adjustment among members of a community. For example, certain reef inhabitants forage at different times of the day. Competition can even be desirable because it eliminates weak and unsuitable

–Photo: Stephen Frink

Prey/predator relationship.
Pictured: A crab scavenges the remains of a dead fish.

members of a species. This is a key principle of modern biology, as described by Charles Darwin.

PREY/PREDATOR RELATIONSHIPS

The most familiar example of an intrinsic factor in a ecosystem is the prey/predator relationship. The predator population benefits itself by obtaining food from prey populations. The prey population may or may not be adversely affected by predation. In fact, some predators benefit a prey population by eliminating unfit individuals or by reducing intra-species competition for food. These "key" predators, such as sharks, help control the number of many other predators. Without these "key predators," the oceans would be over crowded with dead and diseased fish and marine mammals.

Predators may be valuable for other reasons as well. Sharks do not deserve their bad reputation as human predators; on the contrary, they actually help save human lives. Research has shown that sharks have extraordinary immune systems and may hold the key to medical breakthroughs in cures for cancer, bacterial and viral infections, eye cataracts, even AIDS.

Human activities which interfere with the balance of prey/predator relationships can adversely modify an entire marine ecosystem. *If we kill the prey, we kill the predator.* This is why harvesting of many species has to be monitored. For example, depletion of abalone, which is a most popular food delicacy, can result in starvation of its natural enemies–seals and sea otters.

A reverse ecological imbalance can prove just as catastrophic: *if we kill the predator, the prey proliferates unchecked.* The aforementioned predator/prey relationship between the sea otters and sea urchins in the kelp ecosystem is one example. When the sea otter population is over-hunted, the urchins flourish and destroy the kelp beds which are home to many other marine organisms. Another striking example is the problem of the Crown of Thorns starfish, the prickly, multi-armed *Acanthaster planci,* which periodically ravages coral reefs throughout the Indo-Pacific. As few as fifteen of these starfish can eat an area of coral the size of a football field in just two and a half years. One of the factors which scientists believe promotes Crown of Thorns infestation is the excessive collecting of its natural enemy, the spiraled Pacific Triton shell.

Different defense mechanisms are used by marine creatures to survive.
Pictured: A porcupine fish exposes its spines to ward off its enemies.

Photo: Hillary Viders–

-Photo: Mort and Alese Pechter

The dreaded crown of thorns starfish ravages coral.

Defense and Attack Mechanisms

Prey organisms have evolved a remarkable repertoire of defense mechanisms, as varied as the attack strategies of the predators. Some fishes hover in dense schools, making it difficult for a predator to select an individual prey. Sedentary fishes, such as pufferfish or porcupine fish, are covered with spines and can inflate themselves, making them a painful meal at best. Electric eels stun their attackers. Some prey, such as alligators, whelks, and turtles, protect themselves with thick or tough skins.

Butterfly fish display a false eye near their tail; predators, aiming for this false eye, end up overshooting the mark as their prey heads off in the opposite direction. Some reef creatures, such as the scorpion fish employ **camouflage**, a collage of coloring which blends perfectly into the backdrop of the reef, making them almost impossible to find. Other creatures hide under their protective shell or in the shelter of coral.

Predators also have various devices for catching prey. They sometimes hunt in schools, use very keen eyesight, or camouflage themselves until their prey is close enough to attack. Many predators scour their terrain for prey that is weak, sick, crippled, diseased, or dead.

Symbiotic and Parasitic Relationships

Perhaps the most fascinating types of relationships among species are called symbiotic (sym=together, bio=life). SYMBIOSIS is the harmonious association of members of disparate species. There are two types of symbiotic relationships: mutualism and commensalism.

MUTUALISM is a type of species interaction in which both partners benefit. An example of mutualism is the clownfish which lives in the sea anemone. The sea anemone provides shelter for the clownfish, which in turn attracts food for its host. In COMMENSALISM, one partner benefits from the cohabitation, and the other neither benefits nor is harmed. An example of commensalism is the Echeneis remora which attaches itself to a whale, shark, turtle, or

Some fish school in large numbers to avoid being singled out by predators.

–Photo: Keith Ibsen

Some marine creatures, like this scorpion fish, use camouflage to defend itself against predators.

Photo: Stephen Frink–

Marine creatures such as crabs have protective shells for defense.

Photo: Keith Ibsen–

large fish by using its large oval suction disc, (a modified dorsal fin) and feeds on fragments of food which drift free as its transporting host feeds. Another example is that of barnacles living on the jawbones and outer covering of some whales. The barnacles benefit by having a safe place to live and a steady supply of plankton (on which the whale feeds). The whale derives neither benefit nor harm from the barnacles.

Scarcity of food and predation are not the only factors which regulate populations. Parasites also play a key role in ecosystems. In a PARASITIC RELATIONSHIP, the host is harmed, impoverished, or killed by its guest. An example of parasitism is the attachment of an isopod to the skin of fish.

REPRODUCTION AND RECRUITMENT

Essential to the survival of any ecosystem is the reproduction and recruitment of its species. One of Charles Darwin's central contentions was that all creatures are potentially able to produce more offspring than their environment could possibly support. We know, however, that in most species, that is not the case. Competition among organisms is a vital aspect of an ecosystem, and one that can limit the presence or size of a given species population. Marine

Mutualism is a symbiotic relationship in which both partners benefit. Pictured: A clownfish in an anemone.

—Photo: Stephen Frink

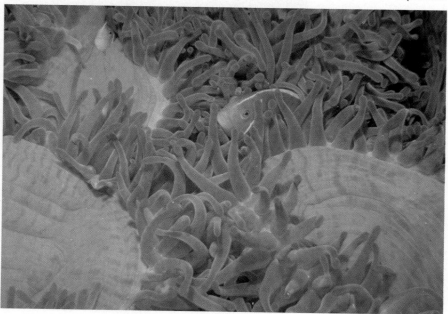

inhabitants, therefore, present a wide spectrum of mating strategies and reproductive mechanisms, both sexual and asexual. An example of reproductive strategy is seen in species whose population rises to enormous heights and then, for various reasons, collapses. Some marine animals, on the other hand, such as whales, produce only a few offspring, but each one has a high chance of survival.

Finding a mate in the vast, three dimensional world of water, often under conditions of little or no light, can be a formidable task. Some marine animals, therefore, use chemical means of finding and attracting members of the opposite sex. Others, such as angler fish, solve this problem a different way. The males are small and permanently attach themselves to the females. Several species of coral reef fishes which are normally widely dispersed, such as groupers, form annual spawning sites where both eggs and sperm are produced simultaneously. Many species of fish and shellfish use wetlands as their protected spawning ground.

–Photo: Stephen Frink

Commensualism is a symbiotic relationship in which one partner benefits and the other neither benefits nor is harmed. Pictured: A remora hitching a ride on a shark.

Photo: Stephen Frink—

In a parasitic relationship, the host is harmed, impoverished, or killed.
Pictured: An isopod attached to the skin of a coney.

Reproduction is essential for marine species to survive. Pictured: Sea turtles mating

—Photo: Stephen Frink

RECRUITMENT is defined as the addition of members to a population. However, recruitment is a subjective term which can mean the number of individuals which reach a certain size for commercial use, the size that reproduces, or the size that makes the individual recognizable within the species. One of the most dramatic problems today, however, is the increase in low fecundity and high mortality rates of many marine species, particularly those with low recruitment, which scientists believe is due to human-induced pollution, habitat destruction, and excessive hunting and harvesting of marine species (see section on "Decline and Depletion of Fish Populations," see page 132).

TYPES OF MARINE HABITATS

"The coastal zone may be the single most important portion of our planet. The loss of its biodiversity may have repercussions far beyond our worst fears." **– G. Carleton Ray, University of Virginia**

Each geographic area, from the kelp forests of California to the coral reefs of the Caribbean and the Indo-Pacific, to the Great Lakes Basin, to the North Atlantic Seaboard, to the great expanse of polar waters, has its own topography and complex array of marine organisms. Yet all these diverse creatures co-exist and play a role in specific ecosystems.

PELAGIC, BENTHIC, AND OFFSHORE ECOSYSTEMS

Within a marine environment, there are PELAGIC and BENTHIC inhabitants. Organisms in pelagic communities live in the water column, such as PLANKTON (drifters), or NEKTON (swimmers). Pelagic life can be found in NERITIC waters (shallow coastal waters) or OCEANIC waters (waters over continental shelves).

BENTHIC communities live within, upon, or associated with the bottom. The widest spectrum of marine species exists in benthic communities, especially in coral reefs.

Due largely to the inaccessibility of all but the shallower areas of the oceans, we know relatively little about open ocean and benthic ecosystems. The accelerating study of these extreme marine environments regularly reveals new and unexpected aspects of life in the oceans. In the last two decades, oil company scientists, for example, have found several dozen new species of plants and animals while performing surveys off the coast of California. So little is known about the animals and plants of the sea floor that scientists were unable to say whether any of the new species were rare or unusual, only that no one had ever seen them before.

–Photo: Stephen Frink

Sharks are an example of pelagic marine creatures.

UPWELLING

The most widely distributed ecosystem in marine waters is that of the open water itself. These open waters are much different from coastal waters. The most obvious difference is that the plants and animals of pelagic waters have no substrate or bottom to which they may attach themselves. Instead they swim or drift unattached.

There are more subtle differences as well. These offshore waters do not receive nutrients from land runoff and rivers. As a result, open ocean waters are generally less productive than the nutrient-rich waters near the shore.

The basis of the open ocean's productivity, as elsewhere in the marine environment, is the great mass of microscopic phytoplankton, the many billions of tiny plants that drift in the upper waters of the oceans and use the energy of the sun to produce carbohydrates. Some of the phytoplankton are eaten by zooplankton, animals not quite as tiny as the phytoplankton.

Some of the zooplankton are the larval or juvenile forms of much larger animals such as crabs, sardines, mackerels, and anchovies. When the larval or juvenile forms of these organisms mature, they join the ranks of the nekton, the free-swimming creatures of the sea, or settle to the bottom to spend the rest of their lives attached to or burrowed into the sea floor. When a member of the nekton dies, it joins the slow rain of other dead animals, plants, and wastes that drifts gradually down to the bottom of the sea. Some of this DETRITUS (detrit=wear off) is consumed as it falls to the bottom. The detritus that reaches the bottom becomes available to benthic animals.

Where the physical and topographical conditions are right, the deposits of detritus may be lifted to sunny surface waters by UPWELLING of bottom waters.

Upwelling occurs predominantly along the west coasts of the continents, where nearshore surface water is pushed offshore by prevailing winds of the rotation of the earth. The displaced nearshore water is replaced by the upwelling deep water. Compared to the surrounding water, the water rising to the surface in an upwelling is cold and nutrient-rich. The combination of sunlight, favorable temperature, and nourishment results in an exceptionally large growth of phytoplankton. In turn, the phytoplankton provides a rich food source for zooplankton, which is then passed on to larger animals. Upwellings, therefore, are some of the most productive marine ecosystems, supporting large populations of marine plants and animals.

A benthic ecosystem

Photo: Stephen Frink–

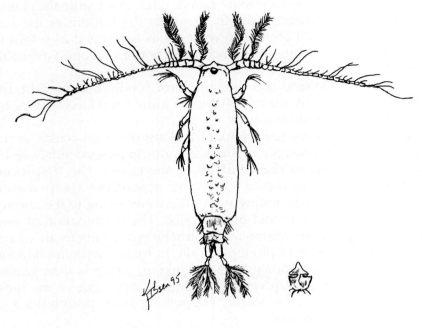

–Illustration K. Ibsen

Figure 7– Comparison of Phytoplankton and Zooplankton

HYDROTHERMAL VENTS

Only very recently did scientists discover a series of self-contained ecosystems deep in the lightless zones of the ocean off the west coast of the United States. The food web in these ecosystems is based not on plants but on certain bacteria that use the energy found in hydrogen sulfide, a chemical poisonous to most animals. No photosynthetic plants are found in these sea bottom-based systems, and none of the animals living there need to scavenge on the remains of plants or animals drifting down from the surface.

The hydrogen sulfide is dissolved in the hot water which flows from springs at fissures and vents in the sea floor. Bacteria get energy from the hydrogen sulfide by oxidizing it. They then use this energy to convert the carbon in carbon dioxide in the sea water into organic carbon. Because the water from the hydrothermal (hydro=water, therm=heat) vents is so rich in hydrogen sulfide, the bacteria are able to form dense masses in and around the hot water vents. Despite their microscopic size, they are the primary producers in the food web of the vent ecosystem, a role normally played by plants. Through the food chain, they support a wide variety of other organisms.

Organisms around the hot springs sometimes take bizarre forms. Among the animals found at the springs are huge clams of a species formerly unknown to science, measuring up to a foot or more in length and having pink flesh. Other visually striking inhabitants of the hydrothermal vent ecosystem are giant red worms up to eight feet in length which live in white parchment-like tubes and a formerly unknown species of mussels. Because their food supply is 300 to 500 times more concentrated than that of the surrounding waters, vent communities are rich in life. But when a vent stops flowing, the supply of hydrogen sulfide is cut off, and the life of the vent ends.

Figure 8 – Shifting Layers of Water and Nutrients During Upwelling

Illustration K. Ibsen –

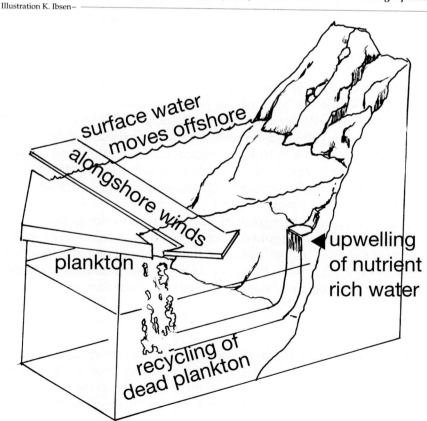

Photo: Jose Azel Aurora–

Okefenokee Georgia swamp

SEAMOUNTS

In many areas of the oceans, the sea floor rises in peaks toward the surface. When the tops of the peaks are above the water, they are called islands. Where they only approach the surface but do not break it they are called seamounts. Seamounts exist in parts of the oceans and usually have little ecological effect. Sometimes, however, the seamounts rise high enough to reach the euphotic zone. When this occurs, they support rich ecosystems unlike those of the deeper surrounding areas.

Commercial and recreational fishermen have long recognized the importance of seamounts. Seamount ecosystems often support important fisheries as fish gather at the seamounts for the food and shelter to be found there. In recent years, seamounts have garnered the attention of marine biologists and oceanographers.

WETLANDS

Wetlands are among the more biologically productive ecosystems in the world; they rival agricultural lands. According to the U.S. Fish and Wildlife Service, wetlands are defined as areas where water is the primary factor controlling the environment and the

*Okefenokee swamp
at midnight*

–Photo: Jose Azel Aurora

associated plant and animal life. These transitional habitats occur between upland and aquatic environments where the water table is at or near the surface of the land, or where the land is covered by shallow water that may be up to six feet deep. Ninety percent of the wetlands in the U.S. are freshwater systems. Scientists have divided wetland systems into five major categories:

1. marine–associated with the ocean;
2. estuarine–associated with estuaries, i.e., transitory between salt and fresh waters;
3. lacustrine–associated with lakes;
4. riverine–associated with rivers; and
5. palustrine–marshes, swamps and bogs.

Wetlands play many roles in preserving the stability of the global biosphere. In a recent study by the Department of Transportation (DOT), ten major uses were identified:

1. to control floodwaters;
2. to recharge groundwater;
3. to filter out pollutants;
4. to provide a habitat for waterfowl;
5. to provide a habitat for marine wildlife;
6. to support fisheries;
7. to provide a sanctuary for rare and endangered species;
8. to provide areas for recreation;
9. to provide areas for education; and
10. to provide areas for aesthetic enjoyment.

The high productivity of wetlands is due to several factors, primarily the ability of its plants to capture the sun's energy via photosynthesis and store it as chemical energy. Wetlands productivity is also a function of both the grazing and detritus food chains. Wetlands are also extremely efficient at recycling energy. For example, wetlands contribute greatly to the cycling of nitrogen; all green plants require nitrogen, but they cannot use the gaseous form in air. Microorganisms in wetlands soil convert nitrogen gas into a form which plants can assimilate. On the other end of the spectrum, harmful and excessive nitrogen compounds which are generated from agricultural and industrial pollution (such as ammonia compounds in fertilizer) can be converted back into harmless nitrogen gas by the bacteria which thrive in wetlands.

In addition, wetlands are "pulsed," or energy-subsidized ecosystems, shaped by the periodic changes in water level. Periodic flooding washes in nutrients, which are then made more accessible when the water level subsides. The flux of the water level, or "pulsing," also provides the critical component oxygen. In both saltwater and freshwater tidal wetlands, the tidal cycle is the driving energy force. (Note: Freshwater wetlands are discussed further in Chapter 8)

SALT MARSHES

Salt marshes, tidal wetlands found in temperate zones, play a key role in the productivity of bay and nearshore waters by generating large amounts of plant material that provides food for shellfishes and fishes, reptiles, birds, marine mammals, and humans. Marshes range from six inches to three or more feet in depth and sometimes begin as shallow lakes and depressions.

Typically, a salt marsh begins with submerged grasses that colonize shallow mud flats. As the grasses capture sediments, they build the shallow water bottom up until it is periodically exposed at low tide. After a while, cordgrasses begin to colonize the bare sediments. As they grow, the cordgrasses slow the flow of the tides, thereby catching sediment and building the marsh upwards. After a point, the weight of the accumulated sediments may cause the marsh to subside. Unless a marsh then receives sediments from a river or runoff from adjacent land, it may continue to sink and become open water again.

Salt marshes are very productive plant factories. Some studies suggest that salt marshes produce more plant material than fertilized agricultural fields. As plants die, fungi, bacteria, protozoa, and other microorganisms break down the plant material, making it available to other creatures from snails and crabs to young fishes and shrimp. These smaller creatures, in turn, are consumed by larger animals from fishes to birds to raccoons. (see photo page 62)

MANGROVES

Mangroves are extremely fertile ecosystems found on about seventy-five percent of the world's tropical shores. In mangrove forests, the line between land and ocean is blurred considerably. Here terrestrial plants have invaded the salt water environment very successfully and created an environment which supports a broad array of animals. Like esturarine and wetland ecosystems, mangroves provide a rich source of food for commercially important fish and shellfish and are also valuable as nursery grounds for juvenile fish, shellfish, sea turtles, and aquatic mammals.

Several abiotic factors affect the distribution of mangrove forest. Mangroves, first of all, grow best where annual average temperatures are above 66°F. Second, mangroves do not develop well in freshwater where they cannot compete successfully with other plants. Third, the fluctuation of the tides alternately wets and dries the soil in which mangroves grow. Lastly, mangroves can grow on sand, rock, and peat, but flourish in fine-grained anaerobic (an=without, aer=air) soil composed of silt, clay, and organic matter.

The key to productivity in mangrove forests and adjacent estuarine and marine habitats is leaf litter. Bacteria, fungi, and other microorganisms cover fallen mangrove leaves, serving as a rich source of protein for shrimps, crabs, clams, and some species of fish. These microorganisms break up the leaf litter, making it edible for still other creatures. Larger fish, turtles, dolphins, and birds, in turn, feed on these creatures.

Mangroves are some of the planet's most diverse and prolific ecosystems.

– Photo: Stephen Frink

Salt marshes are extremely productive ecosystems.

Photo: Pete Nawrocky –

Other plants also contribute to the productivity of mangrove communities. Algae attach to the roots of mangroves and often cover the sediment around the roots. Various species of phytoplankton contribute to the primary productivity of mangrove communities. Turtle and manatee grass also occur in mangrove areas. In less saline waters, other species, known as shoalgrass and widgeongrass, grow.

Although the number of plant species in a mangrove community is small, the diversity of animal species is great. In part, this is because mangrove communities are transitional between marine and terrestrial ecosystems. Such transitional communities, called ECOTONES, have a greater diversity than the communities lying on either side of them. They may be inhabited by species from each of the flanking communities, as well as species not found in either.

Figure 9 – Mangrove Ecosystem

–Illustration: Keith Ibsen

The mangrove's branches and prop roots provide refuge, substrates, and food for an enormous array of shellfish, fish, birds, sea turtles, dolphins, and other creatures. The mangrove seagrasses trap nutrients for these inhabitants, while filtering out pollution and preventing coastal erosion from stormwaters. (See Figure 9)

Because mangrove forests provide both nesting sites and food sources, they attract a much more diverse community of creatures than do other types of coastal ecosystems. In Florida, over 180 species of birds use mangroves for feeding, nesting, and roosting. Green, Kemp's ridley and loggerhead sea turtles, manatees, and bottlenose dolphins are some of the large mangrove inhabitants. Even rare and endangered species, such as the American crocodile and the Florida panther, have been seen in channels adjacent to mangrove communities.

Estuaries

Estuaries are regions where freshwater rivers carrying fertile silt meet ocean tides. Estuaries trap pollutants, thus protecting the open ocean. Because they contain both land and ocean nutrients, they support a complex diversity of life forms, from protozoa to fish-eating mammals and other forms of wildlife. Since salt water is more dense than fresh water, sea water may actually form a salt "wedge" that moves up an estuary below fresh water emptying from a river into the estuary. The location of a salt wedge is determined by tides, wind, and the level of fresh water flowing from a river. The salt wedges in some estuaries are very clearly defined and fluctuate with the ebb and flow of tides and with changes in river flow.

Estuaries and their associated marshes and mangrove forests profoundly influence the lives of many animal species. In one of the premier fishing areas of the world, the continental shelf of the eastern U.S., at least three quarters of fished species spend a portion of their life cycles in estuaries. Yet these prime areas are being degraded so rapidly that entire fish populations are being eliminated. Of 80,000 square kilometers of U.S. esturarine waters, one third is now closed to shellfishing because of habitat destruction.

Seagrass Beds

Like mangroves, seagrasses are adapted to life in the marine environment and serve as nurseries for many species of fish and shellfish. But unlike mangroves, seagrasses are also found in northern and southern waters. Seagrasses grow best in protected areas such as bays and lagoons where waves are not great, although currents may be strong. Seagrasses can tolerate very salty or relatively fresh water and grow best where sediments are fine. As plants, seagrasses require sunlight and cannot survive in turbid waters. Most seagrass resident creatures cannot eat seagrasses directly. Rather, they must wait until fungi and bacteria have broken down the plants, which form a dense mat on the bottom.

Seagrasses in both temperate and tropical regions have not fared well in recent times. Human activities in the Tampa Bay area, for example, have reduced seagrasses there by eighty percent. More recently, a massive die-off of seagrasses in Florida Bay threatens important nursery grounds for commercially valuable species such as pink shrimp and other vulnerable species. In the Chesapeake Bay, submerged aquatic vegetation has declined by sixty percent due to the discharge of pollutants into the Bay and to sewage and silt which is stirred up by dredging. This has contributed not only to a dramatic decline in waterfowl, but also to decreasing catches of commercially valuable fish and shellfish.

Coral Reefs

Coral reefs are prized for their diversity, water clarity, and spectacular vistas. Because many tourists and sports enthusiasts want to enjoy tropical waters, special emphasis should be given to these most fragile and valuable marine environments. Most people would agree that the coral reef is the epitome of underwater habitats for color, diversity, and beauty. When vacationers envision their aquatic activities, they typically imagine clear tropical water with exotic fish and vibrant corals. It is important that environmental educators and guides emphasize that these colorful vistas are alive, sensitive, and fragile.

Corals, primitive colonial animals, have been building some of the richest and most diverse communities on earth for over 200 million years. Reef ecosystems may be comprised of up to 9,000 species of plants, invertebrates, and vertebrates, including fish, worms, mollusks, algae, sponges,and urchins. (See Figure 10) Ichthyologists (zoologists specializing in the study of fishes) believe that a third of all the species of fish in the world live in coral reefs. Because of this abundance, coral reefs are important economically, scientifically, and aesthetically. They are the source of food, new medical discoveries, recreations, and jobs for millions of people in many nations, all of which translates into billions of dollars in revenue worldwide. In many areas of the world, coral reefs also provide a crucial protective barrier against the sea, allowing adjacent land habitation and cultivation.

Corals live within a very narrow range of conditions. Reef-building corals are usually dependent on warm water (above 68°F, or 20°C) that is sediment and pollution free and has a normal ocean salinity (approximately 30-40 parts of salt per thousand parts of water).

Reef-building, or hermatypic (herm=mound) corals grow best at depths shallower than 75 feet (25m). Corals require sunlight and warm water because their inner tissues contain ZOOXANTHEL-LAE (microscopic algae) that depend on light for photosynthesis. Through this process, these symbiotic algae produce food for the host coral and provide assistance with the calcification of coral skeletons.

–Photo: Keith Ibsen

Coral reefs are prized for their aesthetic, ecological and scientific value.

Photo: Keith Ibsen—

The basic unit of a coral reef is the coral polyp.

The basic functional unit of any coral, whether hard or soft, is the polyp (polyp=many-footed). Smaller in size than an aspirin, a polyp's soft, tube-like body has (in most cases) a fringed mouth with many small tentacles that sting for defense or for capturing food. Coral colonies often begin with a single larva that floats as part of the zooplankton for several weeks until it finds a suitable substrate onto which it may settle. After settling, the larva becomes an adult polyp and builds a limestone shelter around itself. In time, the polyp produces a bud which becomes another polyp, also secreting a limestone skeleton about itself. This process continues until a colony of thousands of individual polyps may resemble boulders, staghorns, or fans.

Coral reefs are built up very slowly—in most areas, at a rate of only .39 inch (one centimeter) a year. Over thousands of years, generations of polyps leave behind their skeletons and gradually raise the coral reef closer to the surface of the water. The reef is composed of layers of accumulated limestone skeletons at the base and living corals on the surface. Many of these coral colonies look amazingly like plants. In fact, not until 1753 were coral structures recognized as being made up of tiny animals. Coral belong to the group of animals (coelenterates) that includes anemones, sea fans, jellyfish, and

Portuguese man-o-war, all of which have tiny stinging cells for capture of prey and for defense.

TYPES OF CORAL REEFS

There are many kinds of coral reefs, and conservation-minded people should visit and explore each one to gain a better understanding.

1. FRINGING REEFS are found next to the shore where there is suitable rocky substrate in one to twenty feet of water.
2. BARRIER REEFS are separated from the land by lagoons. Barrier reefs exhibit discrete zones in which different species of corals dominate.
3. An ATOLL is a U-shaped or circular reef surrounding a lagoon. Atolls begin as fringing reefs around volcanic islands. As a volcanic island subsides, the reefs become extensive barrier reefs. Finally, when the volcanic island becomes totally submerged, the reefs are left enclosing a lagoon instead of an island. Coral atolls are frequently found along mid-oceanic ridges that are seismically active.
4. PATCH REEFS may occur behind or in front of larger reefs, such as fringing reefs or barrier reefs. Patch reefs are generally formed by a few isolated boulders or coral. In areas of volcanic activity, deep patch reefs form lava flows.

Although the living corals are only on the surface of the reef mass, they are the essential organism for the maintenance of the coral reef ecosystem. It is important to remember that a reef is not one particular kind of creature—*it is an entire way of life.* A reduction of only two percent in live coral cover can significantly reduce both the number and biological diversity of the coral reef inhabitants. The resistance of a reef community to environmental stress is dependent upon its living coral surface.

Because the reefs are complex and delicate, they are easily disturbed by changes in their environment. Coral reefs worldwide are being damaged at such an alarming rate that scientists are speculating as to whether the damage can ever be reversed.

The decline in coral reefs is a very complex problem with many causes, both natural and human. Since the beginning of time, coral reefs have been subjected to natural disruptive forces such as sea level, temperature and salinity fluctuations, storms and hurricanes, and sediment from bioerosion of the substrata by fish, worms, bacteria, algae, and mollusks.

Historically, the reef's inherent kinetic and chemical energy has usually enabled it to regenerate, but when disruptive forces of nature are compounded by acute human intrusions, the combination may override the reef's ability to recover. Any substantial decrease in the clarity of the water can rob corals of the intense sunlight they require for growth. Pollution and debris are highly destructive to coral. Runoff of sediments from land can also overwhelm the ability of coral polyps to slough off massive sediment. Heavy use of high tech fishing gear, such as drift nets, can severely reduce many fish populations, and irresponsible commercial and recreational activities can damage coral reefs representing hundreds of years of growth by thousands of organisms.

Figure 10– Coral Reef Ecosystem

Illustration: Keith Ibsen–

HARD BOTTOM HABITATS

In the Atlantic, off the southeastern coast of the United States, the broad sandy, mostly barren continental shelf is irregularly broken by banks of rocky outcroppings covered with plants and animals. Off the shores of Georgia and South and North Carolina, such hardground banks occur on the inner, middle, and outer continental shelf at depths of 50-80 feet, 100-130 feet, and 200-330 feet, respectively. Some of these banks may rise only a few feet above the bottom and may be easily covered by shifting sand. Others may have a total vertical relief of 20 feet or more, and include overhanging ledges, caves, burrows, and sandy rock-littered troughs. These rocky outcroppings, which are often called hard bottoms, provide firm surfaces to which plants and animals can attach themselves.

Different communities have developed on the various hard bottoms scattered off the southeastern U.S. Outer and middle shelf hard bottoms, for example, are regularly exposed to the warmer waters of the Gulf Stream. Thus, warm water species of fish and coral, classified as STENOTHERMAL (steno=narrow, therm=heat) because they are able to live only within a very narrow range of temperatures, inhabit these outer and middle shelf hard bottoms. Inner shelf hard bottoms, on the other hand, which are exposed to colder inshore waters and runoff of freshwater from the land, experience wider fluctuations in water temperature and salinity. Species on these hard bottoms are generally EURYTHERMAL (eury=broad) and EURYHALINE (hal=salt).

The cold waters of the North support a great abundance of species, but lack the diversity of species found in tropical waters. Rock bottom and cold water ecosystems are particularly valuable for their variety of algae and seaweeds, which provide a nutrient rich shelter for many marine organisms, including over 100 species of fish. Algae and seaweeds supply humans with compounds that contribute to hundreds of end-products, such as waxes, polishes, soaps, and food stabilizers.

KELP FORESTS

On the west coast of the United States, from the Mexican border to Alaska and the Aleutian Islands, kelp beds are sometimes found in waters shallow enough for light to penetrate to the bottom. These kelp beds shelter millions of individual animals of hundreds of species. Different kinds of kelp are found on the east coast, but these species do not form the giant beds characteristic of the west

coast forms. Kelp forests are economically as well as environmentally valuable, as harvesting of kelp yields many products, from stabilizers in ice cream to a slipping agent for oil-well drilling mud.

Kelp beds are similar to forests and jungles. Kelp plants, like
trees, provide shelter, roosting areas, food, and substrate for a wide
variety of animals. (See Figure 11) Fish swim amid the kelp fronds.
Shrimps and crabs forage for food around the plants.

When scientists refer to kelp, they are actually referring to several different species of large, fast growing marine algae. There are
approximately forty species of kelp along the California, Oregon,
and Washington coastline. Kelp beds off southern California, for
instance, are primarily composed of giant kelp *(Macrocystis)*, while
bull kelp *(Nereocystis)* is dominant from Monterey to Alaska.

As in other ecosystems, one finds zonation in kelp beds. Instead
of zones based on the tidal levels of water, kelp beds are divided

–Photo: USC Sea Grant Program

Garibaldi in a kelp forest.

into the canopy habitat where the kelp fronds are the thickest, from the surface down to about ten feet, the mid-kelp habitat located between the canopy and the area just above the bottom, and the holdfast region, located on the bottom itself.

The three kelp zones are interdependent, and many animals are found in all three. The canopy habitat provides shelter and food sources for juvenile fishes. Among the most significant of these are kelp bass, topsmelt, kelp surfperch, blacksmith, and olive rockfish. Scientists have estimated that as many as 90,000 juvenile fishes may

Figure 11–Kelp Forest Ecosystem

–Illustration: Keith Ibsen

hide in a single acre of kelp. Many adult fish also find food and shelter in the kelp canopy.

Like the canopy, the mid-kelp habitat is used by many adult and juvenile fish. In general, animal life in the mid-kelp region is not so varied or concentrated as in the canopy zone or in the bottom zone, the holdfast region.

The holdfast region, characterized by tangled holdfasts, rocky nooks and crannies, and many inhabitants, is the most complex of the kelp habitats. Many relatively short algae are found at the bottom of kelp beds where they form broad undulating understories and provide additional food and shelter for the bottom community. Thousands of species of animals are found at the bottom, some of the most characteristic being crabs, sea stars, rays, kelp bass, sheepshead, and the nearly ubiquitous sea urchin.

Although kelp forests may appear hardy, they are vulnerable to natural and human intrusions. Lack of sunlight, high seas, and changes in water temperature can have adverse effects. Rough winter storms, and the seasonal warming of "El Nino," for example, have destroyed many of the kelp beds off California, and attempts by scientists to transplant and reattach kelp have met with varying results.

Additionally, there are two known species of sea urchins which attack kelp beds the *Strongylocentrotus franciscanus* and *S. purpuratus.* Some scientists speculate that the proliferation of these deadly sea urchins is fostered by dissolved organic matter from sewage. Since sea otters are the natural predator of sea urchins, where there are healthy sea otter populations, sea urchin populations are kept in check and the kelp beds can flourish.

A more prevalent theory is that increased turbidity and sludge on the bottom inhibits recruitment, plants are not replaced when lost due to storms, and urchins then start moving about eating what is left, further reducing recruitment and survival. In a report published by the U.S. Fish and Wildlife Service, as sludge discharge off Southern California declined, kelp increased independent of urchins.[1]

NEARSHORE TEMPERATE ECOSYSTEMS

The tides ebb and flow over many different surfaces, ranging from the muds of quiet coastal lagoons and the sands of popular bathing beaches to the rocky cliffs and boulder beaches of the shores of Maine and the West Coast. (See Figure 4) Each of the surfaces, or substrates, presents different problems to the creatures living on them. In some instances, the problem is how to hold on to

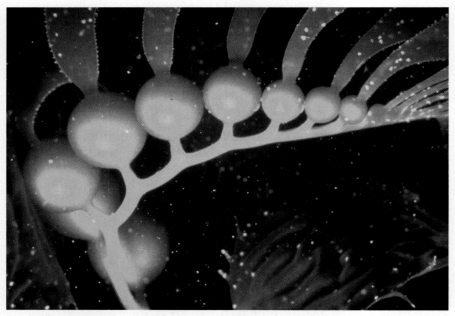

Photo: USC Sea Grant Program–

Kelp pods

the substrate while being continually pounded by waves. In others, it is how to get food or to find mates for reproduction. Nature has found a variety of solutions to each of these problems.

The area between the high tide line and the low tide line is called the intertidal zone. The challenges of living in the intertidal zone are enormous. When the tides go out, organisms living in the intertidal zone are literally left high and dry. As the tides change, waves often smash across the intertidal with great force, threatening to wash away or crush anything not firmly attached or buried. Variations in heat and light also cause stress to the organisms of the intertidal zone.

In the temperate regions of the United States, the two most extensive intertidal ecosystems are the rocky intertidal and the sandy intertidal. Even though rocky and sandy intertidal habitats are often found immediately adjacent to one another, they share virtually no species between them, a mark of how different these two types of ecosystems are. Beyond the intertidal zone are coastal waters that host very productive ecosystems. Shallow parts of the coastal seas are home to vast aggregations of marine life ranging from the commercially important flounders and rockfishes to seals and seabirds.

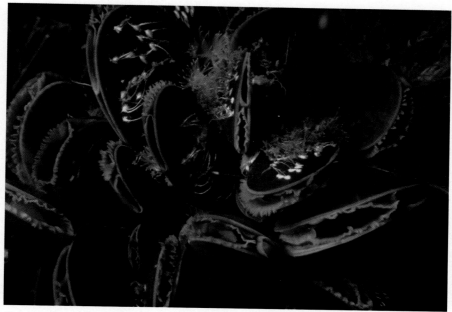

Photo: Pete Nawrocky –

Mussels in a rocky intertidal zone.

The high level of nutrients is the primary factor that distinguishes these temperate coastal ecosystems from others. Because of the nutrients washed into the sea from the land by rain and rivers, temperate coastal ecosystems are naturally highly productive. Their shallowness and temperature enhance this productivity; the temperature of the water is not so cold that it causes stress to marine creatures, nor is it so warm that it interferes with production.

SANDY INTERTIDAL

The intertidal habitat most visited by people is the sandy intertidal. This is the beach between the high and low tide marks. Visitors sometimes see beach crabs and a few other animals such as sand fleas and beach hoppers, but very few people are aware of the sometimes vast numbers of animals which may be hiding beneath their feet. For example, scientists investigating a central California beach determined that at the height of abundance, there were over 3,400 invertebrate animals beneath a one square yard of sandy beach. Most of these animals were tiny crabs and worms; there were some clams and shrimps.

Sandy beaches are subjected to severe and rapid changes in temperature and moisture. Sea water reflects, diffuses, and absorbs the energy of the sun. As the tide goes out, the sand loses the protective cover of water and the exposed sand heats up under the direct rays of the sun. At the same time, the moisture content of the uncovered beach begins to drop, due to both the removal of the overlying water and the effects of evaporation. The animals in this habitat protect themselves from the heat and drying of the sun by burrowing into the sand. Most animals in the sandy intertidal are found in zones, with the effect of the tides being the most important determinant of zonation. (See Figure 12)

As in other marine ecosystems, the physical environment of sandy intertidal areas changes from day to day and from season to season. But other kinds of changes can be catastrophic. Fierce

Figure 12– Sandy Intertidal Ecosystem

–Illustration: Keith Ibsen

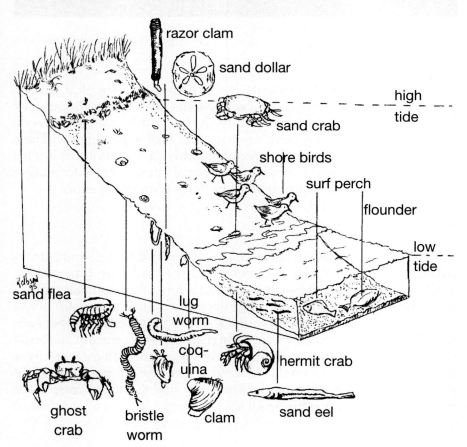

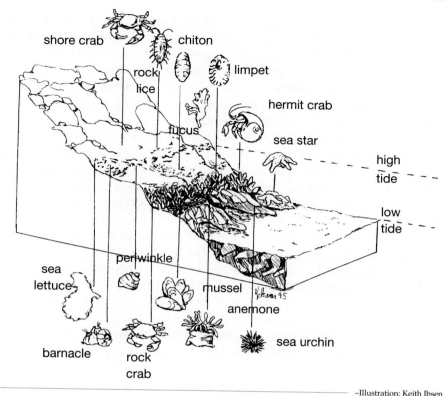

Figure 13– Rocky Intertidal Ecosystem

–Illustration: Keith Ibsen

storms remove large tracts of sand, and imprudent coastal develop-ment increases the natural instability of sandy beaches, leading to their disappearance.

ROCKY INTERTIDAL

The rocky intertidal is a particularly challenging environment for ocean life. Yet rocky coasts are the homes of myriad small crea-tures such as mussels, barnacles, snails, sea stars, crabs, anemones, and a host of marine algae like sea lettuce, purple laver, and feather boa kelp.

A person who looks at a rocky intertidal area for any length of time will soon be struck by the zonation of the plants and animals found there. (See Figure 13) In this environment, as the tides rise and fall and the waves strike and splash the rocks, the tolerance of the creatures is tested. Organisms that can survive prolonged expo-sure to air are found in the uppermost zones. Organisms at the low-ermost zones need to be wet almost constantly.

A significant abiotic factor is the type of substrate. Species that are unable to endure the heat and drying effects of direct sunlight are unlikely to be found on a clean, smooth sandstone substrate. They will hide in cracks and crevices of substrates like granite. A number of biotic factors also affect the distribution of species in the rocky intertidal. There can be intense competition among different species and within species for suitable substrate.

There are positive forces at work in the rocky intertidal as well. The moving water constantly renews nutrients. It also brings in rich supplies of oxygen, flushes away wastes products, and provides protection from predators. These positive factors are sufficient to cause the intertidal zone to be blanketed with dense mats of animals and plants.

The rocky intertidal zone is subject to another set of factors affecting species distribution: the effects of humans. People often come to this area to experience the beauty of the meeting of land and sea. Unknowingly, they sometimes end up trampling well-camouflaged animals and plants and disturbing shy ones. They turn over rocks to see that is under them, causing the death of many living organisms. Sometimes, in the interest of learning more about the intertidal, people collect specimens and remove important keystones in the food chain. These activities, often well intentioned, can quickly deplete large parts of the rocky intertidal.

SUBMARINE CANYONS

The seaward edge of the continental shelf is scored in many places with deep submarine canyons. (See Figure 14) Some of these canyons begin less than one mile offshore in water as shallow as 100 feet and plummet to depths of 14,000 feet or more. The walls of these canyons often rise to more than 3,000 feet, and, in the case of the Great Bahama Canyon, more than 14,000 feet. The animals inhabiting these canyons are totally dependent upon a rain of plant and animal material from surface waters where plant growth takes place.

Generally, the heads of submarine canyons are located on the continental shelf. From there the canyons wind seaward until they terminate on the continental rise. Canyons may stretch more than 200 miles, as does the Bering Sea Canyon, or as little as three miles, as does the Dume Canyon off Southern California. The slope of the canyon floor is greater at the head than elsewhere in the canyon. Often, tributary canyons join a larger canyon near the head and are sometimes suspended several hundred feet above the floor of the main canyon. Submarine valleys, formed from sediment transported down the canyon bottom, may fan out for many miles from the mouth of the canyon.

Because plants cannot grow in these deep canyons, life in the canyons depends upon the plant life in the shallower waters above the canyons. Most of the food in submarine canyons, however, is waste from zooplankton and the carcasses of marine animals. As this food falls from the surface waters, it is consumed by bacteria and detritus feeders. As a result, the amount of available food decreases with depth.

Some food material may spill into the canyons from the continental shelf. Canyons are sometimes richer in food sources than the surrounding continental slope and rise because canyons, like funnels, focus food spilling from the continental shelf. Also, animals migrate with the transfer of food from shallower to deeper waters. Predators in shallow waters may become prey in deeper waters.

Because of the remoteness of submarine canyons, much remains to be learned about the dynamics and distribution of animal popu-

Illustration: Keith Ibsen–

Figure 14– Geographical Distribution of the Submarine Canyons

lations in these areas. Studies off the east coast of the U.S. suggest that canyons have a greater abundance of epibenthic (epi=on) animals, such as anemones and soft corals, than do the continental

slope and rise. This is because canyons offer more hard substrate to which these organisms can attach than do the continental slope or rise. These epifauna (fauna=animals) also depend upon currents to carry food to them.

A variety of animals move about the bottom feeding upon other animals or upon detritus. Among the most active of these animals are red and jonah crabs. Fish inhabitants include the rattail (related to the cod), hake, eel, and flounder. Because submarine canyons have only recently become the focus of increasing scientific study, much remains to be learned about how the canyons are shaped, their productivity, and the interactions of their inhabitants. (See Figure 15)

–Illustration: Keith Ibsen

Figure 15– Submarine Canyons Ecosystem

Nearshore Arctic Ecosystems

North of Alaska and the Aleutian Chain are parts of the Arctic Ocean, the Beaufort and Chuckchi Seas. These cold, shallow waters are the locations of some unusual ecosystems in which one physical factor–ice–is often the overriding determinant of species distribution.

Although sea water has a freezing temperature much lower than that of fresh water, beginning in October, the waters of the Beaufort and Chuckchi Seas annually freeze to a depth of at least six feet and do not become clear again until the following June. The combination of the thick ice with the far northern location, where the sun shines for only a part of the year, means that most of the time the waters of the seas are pitch black. Plants and animals have made unusual adjustments to this fact of Arctic life.

The rivers that empty into Arctic seas have laid down vast deposits of mud as they have flowed down from the Brooks Ranges for thousands of years. In most respects, Arctic muddy bottoms are no different from other soft bottom marine ecosystems except for one factor–the annual ice scouring they receive by the action of wind, waves, and currents which dislodge and drive large ice floes along the bottom of the shallow sea floor.

The first few divers who ventured into the cold and dark waters of the Arctic described the seafloor as flat, featureless, and lifeless. Closer looks, however, have shown otherwise. Within the mud bottom, numerous forms of filter-feeding worms and clams are found. These animals feed on detritus, plankton, and algae, and are in turn the main source of food for cod, char, whitefish, capelin, and smelt.

The ice-gouged sea bottom of the Arctic is an outstanding example of how the physical characteristics of an area shape an ecosystem. Life in the frigid waters of the Arctic is adapted to the conditions of a silt bottom, freezing water, and constant reworking of the sediment by ice.

Antarctica

Antarctica, an island continent, is a study in extremes, it is the coldest, driest, windiest, most remote and most undeveloped region on earth, yet it teems with marine life. The landmass, a glacial cover which has accumulated over 25 million years, accounts for ninety percent of the earth's ice and seventy percent of its fresh water. Antarctica expands to twice its size in winter when the sea freezes, and it contracts in summer, when the ice melts and breaks off as huge floating icebergs.

Photo: Jose Azel/Aurora –
Antarctica Pictured: Adelle penguins on Paulet Island on the Antarctica Peninsula.

Vista of Portage Glacier, Alaska

Photo: Richard Townsend/courtesy of the Center for Marine Conservation–

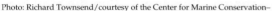

Antarctica is a vital ecosystem which regulates global climate and sea levels. Its ice mass draws heat from the tropics and circulates cold air and water to the northern regions of the earth. Its icy waters absorb large amounts of carbon dioxide from the atmosphere; this helps regulate the earth's average temperatures.

Because of the cold and dryness, Antarctica's landmass has virtually no plant life, but the coastal waters, rich in oxygen, nutrients, and sunlight during the summer months, support a profundity of phytoplankton, which are eaten by krill, which in turn, comprise the base an enormous food web. Antarctica has approximately 85 million penguins, 80 million seabirds, hundreds of whales from a dozen species, half of the earth's seal populations, and 200 species of fish.

In the last few decades, Antarctica has been the center of heated scientific and political debate, as various international interest groups have sought to mine its vast wealth of natural resources–minerals, oil, natural gas, coal and metals. Seven countries have attempted to claim ownership of Antarctica, and twenty six countries operate research stations there. In 1959, an international treaty declared Antarctica a nuclear-free zone dedicated to peaceful pursuit and exchange of scientific information. In 1977, the thirty-nine countries involved in the Antarctic Treaty agreed to ban

Antarctica, Pictured: Icebergs near Yalour Island on the Antarctica Peninsula.

Photo: Jose Azel/Aurora–

mineral mining and development until a treaty regulating these activities was drawn up. Although the Convention on the Regulation of Antarctic Mineral Resources was created in 1988, it has not be ratified by any country, and there is still strong disagreement as to what activities should be allowed.

In May, 1995, Seoul, Korea hosted the most recent Antarctic Treaty Consultative Meeting, which reaffirmed the coalition's disappointment and lack of meaningful progress towards the protection of Antarctica. More than three years after the signing of the Madrid Protocol, a landmark agreement that prohibits mining in Antarctica and set new environmental standards for the region, the Antarctica region is still at risk of serious environmental damage. According to many countries, the Madrid Protocol is essentially a hollow gesture until it is ratified by all twenty-six full members of the Antarctic Treaty. To date, only sixteen have done so. Additionally, Antarctica is threatened by the absence of a liability regime that would cover damages to the Antarctic environment.

FRESH WATER ECOSYSTEMS

"The ecosystems of wilderness areas such as Lake Baikal are priceless and fragile. They can not ever be taken for granted." — **Jim Fowler, Exec. Dir. Wildlife Heritage Trust**

Although fresh water is a critical habitat and component of ecological cycles, less than three percent of earth's water is fresh. Of this small portion, seventy-seven percent is frozen in polar caps and glaciers, twenty-two percent is groundwater, and the remaining small fraction is in lakes, rivers, streams, and ponds. Yet, these fresh water ecosystems and their resources have always been an integral part of the global environment. Water's ability to irrigate crops, for example, was a vital factor in the birth of civilization, as evidenced by the fact that the first ancient societies developed along the banks of major rivers such as the Tigris, the Euphrates, the Nile, and the Indus.

Besides their agricultural value, fresh water ecosystems are home to thousands of species of fish and other wildlife. Scientists classify freshwater ecosystems into three categories:

1. lotic (running water): rivers and streams;
2. lentic (standing water): lakes and ponds; and
3. wetlands: marshes and swamp forests.

RIVERS AND STREAMS

Although small by comparison to the oceans and land masses of the world, rivers are among the most essential natural ecosystems. By some estimates, the animal diversity in rivers is sixty-five percent greater than that of the sea. The Amazon River, for example, which contains between 2,500 and 3,000 species of fish, is believed to be the most diverse ichthyological region in the world.

For millennia, man has depended on rivers to provide drinking water, food, transportation, agriculture, flood control, and an

Fisherman on the Nile River near Kom Ombro in Egypt.

For centuries, the Nile has been a centripetal force in civilization.
Pictured: Farmer using a waterwheel during harvest

avenue for waste disposal. By eroding the land through which they flow and by depositing sediment, rivers help shape terrestrial landscapes and ecosystems which nourish plants, insects, birds, crustaceans, mollusks, and mammals. In effect, rivers have profoundly shaped the course of human history.

Beginning with the downward flow from mountain highlands, and ending in the sea, water in a river system passes through three phases, each one containing different environmental factors, and therefore different ecosystems.

LAKES AND PONDS

Lakes and ponds are bodies of standing fresh water formed when precipitation, runoff from land, or flowing groundwater fills depressions in the earth. These basins are relatively young compared to the ancient history of the earth's oceans. Depending on its size, the lifespan of a pond usually ranges from a few weeks to a few hundred years. According to scientists, standing water ecosystems change with time at rates more or less inversely proportional to their size and depth, and species diversity tends to be relatively low.

Ponds contain a variety of plants. Some plants emerge from the water (cattails, rushes, pickerelweeds, arrowhead), others partially

Debris is a problem in fresh water as well as marine environments.
Pictured: Debris strewn in the Great Lakes

Photo: Michigan Sea Grant —

float on top of the water (water lilies, water-shield), and others are completely submerged underwater (pond weeds, milfoil). These plants serve as food and shelter for the pond's aquatic animals. The pond also has herbivores (planktonic animals, copepods, etc.) and animals which graze directly on plant plankton, such as tadpoles, frogs and toads, and mosquitoes. Some of these creatures are in turn consumed by carnivores (sunfish, bass, pike, turtles, water snakes, etc.).

Lakes are characterized by four distinct zones which provide a variety of habitats and ecological niches for different species:

1. littoral zone–includes the shore and the shallow nutrient-rich waters near the shore. This zone contains a variety of phytoplankton, rooted aquatic plants, and other forms of aquatic life;
2. limnetic zone–open water surface layer in which photosynthesis takes place. The amount of phytoplankton, zooplankton, and fish in the limnetic zone depends on the supply of nutrients;
3. profundal zone–this deep open water area is darker and cooler. It is inhabited by fish and organisms which do not need abundant sunlight; and
4. benthic zone–the bottommost layer of a lake. It is inhabited by decomposers, such as bacteria and fungi, and organisms which feed on detritus.

Most scuba divers who frequent lakes have experienced thermoclines, or layers of water which undergo thermal stratification during winter and summer. During spring and fall, the entire lake approaches a similar temperature throughout and the layers of water are "mixed." This occurrence is often followed by a plankton "bloom."

Primary productivity in a standing-water ecosystem depends on its chemical nature, depth, and the amount of siltation and nutrients present. Depending on their degree of primary net productivity, lakes are classified as EUTROPHIC (eu=well, trophic=nourished), OLIGOTROPHIC (olio=poorly), or MESOTROPHIC (mes=half). Eutrophic lakes contain an abundance of nutrients, phytoplankton, zooplankton, and fish, and tend to be shallow. In warm summer months, the bottom layer of an eutrophic lake is often devoid of dissolved oxygen. Oligotrophic lakes, which have very low primary productivity, have deep, crystal-clear blue or green water and steep banks. Mesotrophic lakes have a medium amount of primary productivity.

In general, the oldest and deepest lakes have the greatest biodiversity. In Lake Baikal, for example, which is believed to be twenty-five million years old, scientists have recorded over 1,500 species of animals and 1,000 plants, at least 1,300 of which are found nowhere else in the world. It is not surprising, therefore, that for millennia man has depended on fresh water environments for drinking water, food, transportation, and agriculture.

Human-Induced Stresses in Fresh Water Ecosystems

Most lakes undergo some natural degree of EUTROPHICATION, that is, physical, chemical, and biological alterations due to nutrients and silt which run off from the surrounding land over time. Excessive eutrophication which results from human intervention, however, is a serious problem. Because of agriculture and urban activities, nitrates and phosphates from human sewage, nutrients from fertilizers and animal wastes, and erosion of topsoil are contaminating lakes and altering their ecosystems at an alarming rate. Eutrophication can kill valuable aquatic species, such as trout, which need cool, clear, oxygen-rich water. Eutrophication also fosters the proliferation of algae and aquatic plants which can entangle swimmers, boaters, and fishers, and which impart a foul taste and odor to the water.

In effect, thousands of fresh water lakes and rivers and their biota are being destroyed faster than they can be protected because of human-induced impacts, particularly pollution. Chesapeake Bay, the Mississippi, the Hudson River, and the Potomac are only a few tragic examples of what pollution can do to the aquatic environment. A three-mile strip along the Niagara River is a case study of human trespass. Besides pollution, species are declining because of overfishing, habitat destruction, and the introduction of alien species. When viewed as components of the earth's biosphere, fresh water ecosystems are vulnerable to impacts in terrestrial ecosystems. For example, deforestation, or removal of trees and woody vegetation, causes a deleterious increase in the flow of streams and rivers, thereby increasing soil erosion and disrupting the natural cycle of water regulation.

Even the Nile River, a source of life for thousands of years, now faces a growing pollution threat from industry, agriculture, and sewage. More than 20,000 factories pour hazardous waste into the Nile. One of the worst pollution sites is Manzala Lake in Northwest

– Photo: Chris Kohl

Diver inspecting the propulsion mechanism of the wooden freighter, the "Conemaugh,"
in Lake Erie, June, 1988.

Diver inspecting the same propulsion mechanism on the "Conemaugh," now encrusted with
zebra mussels, thirteen months later.

Photo: Chris Kohl –

Egypt, which is fed by the Nile and smaller rivers. Manzala Lake used to be one of Egypt's primary sources of fish, but fish production has dropped seventy percent because of pollution from agricultural pesticides, untreated sewage, and industrial waste.

In the U.S. and Europe, the demands of human population and industry on a limited supply of fresh water has seriously degraded water quality, species diversity, and the natural functioning of fresh water ecosystems. Rivers are routinely dammed, channeled, drained, and polluted with salts, silt, nutrients, and toxic waste. Some lakes and rivers have become so inundated with human-induced toxins (such as acid rain) and debris, and depleted of oxygen, that no species of fish can survive at all.

INTRODUCTION OF ALIEN SPECIES

For centuries, alien species (including plankton, barnacles, mollusks, hydroids, worms, algae, fish, and shellfish) were carried from one country to another on hulls of vessels. In modern times, alien species have been transported in the ballast of ships. Since the 1970s, there has been recorded an unprecedented rate of invasion and proliferation of alien species, some of which have created havoc within their host ecosystems, both marine and freshwater. Alien species, such as the sea lamprey, which kills trout and salmon, have no natural predators in the new ecosystem.

An invasion of an alien species which has received much publicity is that of the Zebra mussels in the Great Lakes. Scientists believe that larvae of this salt water mollusk, a native of the Caspian Sea, was transported to North America from Europe aboard a ship that flushed its ballast in Lake St. Clair in the spring of 1988. The Zebra mussel was able to adapt to freshwater conditions and began to multiply and spread rapidly, and by fall of 1989 was growing to thirty thousand mussels per square yard in some areas. The Zebra mussels soon began clogging water intake pipes in Lake St. Clair and Lake Erie, causing a dramatic drop in water supplies. Zebra mussels also pose a serious ecological threat because they eat algae that form the basis of the aquatic food chain, thus altering the food chain and reproductive cycle of spawning fish.

Introduction of alien species also affects divers and scientists as well. Since their arrival, Zebra mussels have be progressively encrusting submerged shipwrecks, the core of Great Lakes diving and maritime archeology. The average female Zebra mussel provides from thirty thousand to forty thousand offspring in one season, so several layers of Zebra mussels, several inches thick, can accumulate within one year. Chris Kohl, a Great Lakes underwater photographer has been documenting the infiltration of these aliens on the Great Lakes shipwrecks, and one of his typical "before" and "after" pictures illustrates the rapidly increasing damage.

RESERVOIRS

Reservoirs are large, deep, artificially-created bodies of standing water. Although reservoirs usually contain fresh water, some contain seawater, such as the marine reservoir in Hong Kong. Reservoirs store water which can be carried by aqueducts to urban areas for drinking. They can also be managed to produce hydroelectric power, or to irrigate land, or to prevent nearby land from flooding. Some reservoirs are even used for recreational activities such as swimming, fishing, and boating.

However, many reservoirs are created at the expense of altering rivers and lakes, along with their valuable ecosystems. In essence, rivers and lakes have been modified in many parts of the world for centuries, but the accelerated development of hydroelectric power over the last few decades may eventually dominate a worldwide landscape of regulated rivers. According to the World Register of Dams, more than 12,000 large dams were built or under construction on major rivers by 1968, and as we near the twenty first century, dam construction continues to escalate. With increasing global pressure to harness power and to obtain flood control, irrigation waters, and urban drinking water supplies, some scientists predict that by the year 2,000, more than 60% of the world's total stream flow will be regulated.

FRESHWATER MARSHES

As mentioned in Chapter 5, wetlands are arguably our planet's most fertile and biologically important ecosystems. Many of the factors and activities present in salt marshes, mangroves, and estuaries also apply to coastal river marshes and forested wetlands between land and rivers and lakes. Freshwater wetlands also include bogs, fens, and many other kinds of more isolated ecotones between land and groundwater.

The favorable combination of sun and water flow determine the function and fecundity of freshwater wetlands. Like their marine counterparts, these ecosystems are invaluable as nursery grounds for fish and shellfish, habitats for migratory and domestic sea birds and assorted marine and terrestrial wildlife, natural filters for pollutants and excessive nutrients, barriers against flood waters, and receptacles for carbon dioxide.

THE FLORIDA EVERGLADES

The Florida Everglades are an exceptionally large and interesting expanse of freshwater marshes, characterized by naturally fluctuating water levels. Sometimes referred to as "a river of grass," the Everglades are compact islands of tropical and subtropical foliage which rise from a vast prairie of tall, sharp-bladed sawgrass. These sprawling marshes, important in maintaining water tables under coastal cities and for providing food and shelter for some of Florida's most valuable natural wildlife, were designated a National Park in 1947.

Over the last few decades, this freshwater environment, which supports numerous biotic communities, has been intruded upon by human ambition and greed. Draining, siltation, declining water levels due to the construction of canals and levees, pollution, and a glut of tourists and hunters reduced the number of species which nest in the Everglades, particularly wading birds (herons, ibises, storks, egrets), by nearly ninety percent. In the 1960s there were plans to build a jet airport that would have destroyed large tracts of swampland which supply much of the Everglades with its surface water. Fortunately, the project was scrapped because of strong opposition by environmentalists. In 1994, a developer who owned several hundred acres in the Everglades announced his intention to drain the swamps and build a large amusement park. This proposal is still being disputed.

In addition to the threat of habitat destruction, the health of an ecosystem often depends on one or more key member species. A case in point is the alligator which inhabits the Florida Everglades. The female alligators dig up mud and grass to make a nest for their young. These "gator holes" become ponds which serve as water reservoirs for many of the local animals (bobcats, raccoons, etc.) during dry spells. Gator holes also provide a refuge for fish and birds. Poachers, concerned only with the short-term profit they can make from selling alligator skins, are hunting this keystone species to near extinction, and by doing so, creating a long-term, possibly irreversible ecological imbalance.

SCARCITY OF DRINKING WATER

Looming over the ecological threat is an even more immediate issue, which is already affecting the lives of millions of people: lack of access to drinkable water and severe water shortages. The rising demands for water for agricultural, industrial, and domestic use (in some regions exacerbated by drought), have in effect made safe drinking water scarce in many developing countries. According to studies conducted by the Natural Resource Defense Council and the Environmental Working Group, more than one in five Americans unknowingly drink tap water which is polluted with feces, lead, radiation, and or other contaminants. The studies, which used EPA statistics for 1993-1994, reported that 53 million Americans drank water that violated EPA safety standards under the clean water act. This number constituted an increase of 7.6 million over the 1991-1992 data.

On a global scale, millions of people die each year from drinking contaminated water, while groundwater, rivers, and lakes continue to be inundated with sewage and industrial pollution. In other instances, abundant water supplies are simply poorly managed and wasted. In the U.S., the vast groundwater supplies of eight Great Plains states have been reduced so much that the water table is falling by about three feet (one meter) a year. In Southern California, drought, coupled with over-consumption, has caused severe water shortages.

Florida Everglades

Photo: Jose Azel/Aurora —

The Florida Everglades supports an enormous array of wildlife. Pictured: A gallinule protecting her eggs.

–Photo: Mort and Alese Pechter

94

<u>NOTES</u>

ENVIRONMENTAL STRESS: NATURAL EFFECTS

"Nature is not a pretty, manicured place maintained for human beings. It is a dynamic continuum, often a violent one."
— Dave Foreman, Founder of Earth First

As evidenced in fresh water environments, marine ecosystems are affected by natural and human-induced changes which cause stress. There are four basic ways in which living organisms in an ecosystem can respond to environmental stress:

1. decrease their birth rate or increase their mortality rate;
2. migrate to a similar but less stressful environment;
3. adapt to the environmental changes through natural selection; or
4. become extinct.

Some forms of stress are gradual (i.e., contamination); others are acute and catastrophic (habitat destruction). To develop solutions to problems in the marine environment, it is necessary to understand the origins and interrelationship of these problems. Therefore, the following two chapters will discuss various kinds of environmental stresses, and their long-range consequences.

Natural disturbances include El Ninos, coral bleaching, red tides, earthquakes, volcanic eruptions, tsunamis, hurricanes and unusually heavy rain, fluctuations in sea water temperatures, extreme tides, erosion of substrate by fish and marine creatures, natural waste, disease, and population explosions of predators.

El Nino

El Nino is an example of a natural stress to the marine environment and an illustration of the interdependence of organisms in a marine ecosystem. El Ninos, referred to by scientists as ENSOs (El Nino-Southern Oscillations), are the periodic warming of the equatorial Pacific Ocean, which is one of the world's most powerful weather makers. El Nino which is spanish for "the child," supposedly received its name because it is most notable at Christmas time. El Nino has been observed on the Peruvian coast for hundreds of years ever since ships first anchored at the Port of Callao and their hull paint was blackened by the sulfides produced from decaying marine vegetation produced during an El Nino.

El Ninos are precipitated by shifting winds that occur offshore of Peru and Ecuador every two to nine years. El Ninos interfere with upwelling, the rise of colder nutrient-rich waters to the surface, resulting in a twenty-fold decrease in algae. Additionally, the abnormally warm sea-surface temperatures associated with El Nino are known to set off chains of atmospheric events that dictate the weather in many parts of the world, including North America. El Nino's reach is so powerful that scientists consider it a reliable indicator of rainfall and corn yields as far away as South Africa.

Although there are theories regarding how these phenomena originate, it is still not known why some El Ninos grow large in size and others do not. El Nino starts when prevailing easterly winds slacken at the Equator, allowing broad but subtle waves of warm water from the western Pacific to flow eastward toward South America.

In 1982-83, a disastrous El Nino killed many people, caused billions of dollars in damage, and took an enormous toll on the environment. Scientists who visited offshore islands in 1982-83 expected to see a large bird population and instead found a few nests with starving juvenile birds left behind. In 1982 there were 8,000 great frigate birds on the islands. After an El Nino had occurred that year, there were fewer than 100. The warm waters brought by El Nino prevented the production of algae which also resulted in record low numbers of anchovies and other fish such as mackerel which feed on tiny fish which in turn feed on the algae. When the food resource disappeared, marine birds and marine animals also suffered large population declines.

Since the disaster of 1982-83, a satellite monitoring system has been set up to alert scientists to atmospheric and oceanic changes in the Pacific Ocean associated with global climactic changes. (See Bermuda Biological Station for Research, see page 236). Data based on satellite tracking suggest that the negative effects of an El Nino event may be long-lasting. A study by oceanographer Dr. Gregg A. Jacobs and six colleagues at the Naval Research Laboratory at the Stennis Space Center near Bay St. Louis, Mississippi found that the warm wave of 1982-83 in effect bounced off the American coasts and was propagated northwestward. This northwestward wave was of the same strength and extent as those seen in the tropics during El Ninos, and it significantly raised sea-surface temperatures in the northwestern Pacific. Although the study has not been conclusive, the researchers believe that this was responsible for aberrant weather patterns, which included record-breaking rains and flooding over the North American continent during the past decade.

Coral Bleaching

Coral bleaching, often following El Nino, is another natural stress. This disease received is name from the skeleton-like, bleached appearance which coral take on when the chlorophyll in their zooxantellae becomes less concentrated or when the zooxantellae is expelled from the coral polyps.

Vast areas of coral bleaching were observed in the Eastern Pacific following the El Nino event of 1982-83. In Costa Rica, half the corals on the Pacific coast reefs died, up to 80% of the corals on the Panama reefs perished, and 95% to 100% of reefs in the Galapagos Islands died.

Incidents of coral bleaching may also occur for other reasons and in parts of the world which are not affected by El Nino, such as the Caribbean. It is believed that coral bleaching can be triggered by one or more different factors: abnormal rise or fall in water temperature, excessive dilution of sea water with fresh water, intense sunlight, or pollution. Global warming is also believed to stimulate coral bleaching, although the exact mechanism is not fully understood by marine biologists. Nor can scientists explain why some species, such as branching corals, are more vulnerable to bleaching than others, or why parts of the same reef may be affected and others not.

Red Tides

Although red tides may be caused by human-induced pollution (particularly sewage effluent and runoff from farms and lawns), red tides have been occurring naturally since biblical times. A red tide results from a rapid increase in the production of DINOFLAGELLATES, one-celled organisms which thrive in water rich in nitrogen and phosphorus. Dinoflagellates literally ravage these nutrients and reproduce or "bloom" so profusely that they cover the water like a red carpet. By using all the oxygen in the water and blocking out the sunlight which plants need, an episode of red tide suffocates marine life and kills off massive amounts of fish.

EARTHQUAKES AND VOLCANIC ERUPTIONS

The earth's crust is broken up into seven large shell-like plates and a dozen or more smaller plates. Scientists believe that these tectonic plates are constantly in motion with respect to one another–they pull away from each other, slide past one another, or converge. When plates are pulled apart, a crack forms between them which fills with hot material rising from the earth's center. These movements, powered by heat from the decay of radioactive chemicals, cause the crustal plates to slide over a partially molten surface. These movements can produce earthquakes, volcanoes, and other changes on the earth's surface, which in turn, can impact neighboring coastal and marine areas.

Volcanic eruptions can cause immense damage to reefs or prevent their formation. In Hawaii, for example, there is little reef growth around the most volcanically active islands, whereas rich reefs develop around older, inactive volcanoes.

Volcanic eruptions can also influence global climactic changes. The 1991 eruption of Mount Pinatubo in the Philippines, for example, exerted a cooling effect on the entire planet by casting a sunlight-reflecting haze of sulfate aerosols into the atmosphere. This is thought to have played a role in upsetting weather patterns.

TSUNAMIS

A TSUNAMI (Japanese for "harbor wave") is an ocean wave caused by sudden motion of a portion of the ocean floor or the shore, such as that caused by an earthquake, volcanic eruption, or landslide. A tsunami that overflows the land is often referred to as "tidal wave," although it has no relation to the tide.

If a volcanic eruption occurs below the surface of the ocean, the escaping gases cause water to be pushed upward in the shape of a dome or mound. The same effect is caused by the sudden rising of a portion of the ocean floor. As this water settles back, it creates a wave that travels at high speed (averaging 500 miles per hour) across the surface of the ocean. Although these waves do not rise above three feet in deep water, they can become extremely large and destructive as they approach shallow water and run up on the shore. Waves over 100 feet have been recorded. Although destructive waves from tsunamis are rare, when they do occur, they can ravage entire coastal, terrestrial, and marine ecosystems. In 1946 an earthquake near the Aleutian Islands caused a tsunami that spread over the entire Pacific, killed 173 people, and destroyed twenty-five million dollars worth of property and miles of coastline. The catastrophic tsunami which resulted when the volcano of Krakatau erupted in 1883 brought 115-foot high waves which annihilated 1000 villages and killed 36,000 people–making it one of the worst natural disasters in history.

HURRICANES

Hurricanes, sometimes also referred to as tropical storms, cyclones, or typhoons, depending on the intensity and geographical location, can wreak havoc in the marine environment. In tropical reef areas, such as the Caribbean and the Pacific, hurricanes periodically decimate large tracts of coral, inundate the water with sand and silt, and demolish miles of coastline. Unlike El Ninos and volcanic eruptions, hurricanes occur frequently. There are several hundred hurricanes a year in the Caribbean alone.

Some scientists believe that hurricanes may have a positive effect on coral reefs, because the faster-growing corals, such as elkhorn and staghorn coral suffer the most, thereby allowing the slower-growing species, like the more robust massive corals, a chance to catch up. The massive corals are kept in balance, however, because branching corals recover and recolonize the reef fairly rapidly.

Corals are not the only organisms impacted by hurricanes. In 1989, after Hurricane Hugo hit the island of St. Croix with winds of 144 miles per hour, thousands of dead fish and invertebrates were washed up on the beaches. After such a natural disaster, fish populations initially decline, but normally recover within a year or so.

NOTES

ENVIRONMENTAL STRESS: HUMAN EFFECTS

"It is only in the most recent and brief period of their tenure, that human beings have developed in sufficient numbers, and acquired enough power, to become one of the most potentially dangerous organisms that the planet has ever hosted."

– John McHale

Humans have always regarded the marine environment as their domain to exploit. However, since World War II, urbanized countries have looked to technology both as a means to harvest resources more quickly and efficiently and as a panacea for the undesirable side effects. Many governments still foolishly believe that no matter how dramatically we deplete the earth's resources and allow world population to multiply, technology will always bail us out.

Many of the human-induced stresses discussed in this chapter, such as sophisticated industrial pollutants, acid rain, and radioactive waste, are relatively new to the natural landscape. Their effects, however, are far-reaching, and very possibly impervious to technological curative efforts.

POLLUTION

Most wastes we discharge into the air, water, and land eventually wind up in the ocean. Twenty-two billion tons of pollutants are dumped into the oceans every year, mostly from land-based and atmospheric sources. The global circulation of ocean currents, the longevity of many pollutants, and the continuity of marine life means that no part of the ocean is exempt from pollution. Additionally, we have to consider the cumulative effects of contaminants, whether DDT, PCBs, or trace metals, and their synergistic interactions. For example–siltation can interact with pollutants, changing their form so that they can be more readily ingested.

The United Nations Group of Experts on Scientific Aspects of Marine Pollution estimates that forty-four percent of marine pollution

Point source pollution. Pictured: Industrial waste spewed through a pipe.

–Photo: Soil Conservation Service

originates as land-based runoff and discharges, thirty-three percent is from atmospheric sources, twelve percent is from boat discharges and spills, ten percent is from deliberately dumped wastes, and one percent is from offshore mineral mining. There are three basic types of land-based pollutants found in the ocean: dissolved nutrients, dissolved toxins, and suspended toxic or non-toxic particles transported by rivers and streams into the ocean. Each of these can be disastrous to marine environmental health and viability.

Scientists and regulators classify water pollution into two broad categories: "point" and "non point" sources. Point sources come from a single point, the mouth of the pipe, for instance, or from a distinct location, such as a factory, ditch, well, vessel, ship yard, or container. Non-point source pollution is indirect, such as storm runoff drains carrying pollutants from streets, construction sites, car exhausts, and farmlands.

SEWAGE

Sewage is a problem wherever there are dense human populations. Sewage can be point or non-point (runoff) pollution. Point pollution occurs in channels such as ditches, canals, and pipes, which release sewage in a specific area. Even in cities and municipalities that have sewage treatment facilities, treated sewage "sludge" may be pumped directly into the sea, dumped from barges, or carried through storm sewers into rivers and streams and then into the ocean. Sewage adds phosphates and nitrates to the water, which cause explosive algae blooms. As these nutrients decompose, they use up the oxygen supply which marine organisms need. Ultimately, when the oxygen is depleted, the result is an anoxic, or "dead zone."

NOTE: Marine creatures also produce a large amount and variety of excrement, but the sea can deal with these natural wastes because they are widely dispersed. On the contrary, human waste materials are usually dumped in high concentrations in small areas, such as from municipal sewage systems.

A dramatic example of water which became anoxic was seen in the New York-New Jersey area in the late 1970s and early 1980s, where nine million metric tons of sludge were being dumped every

Sewage pollution

Photo: EPA–

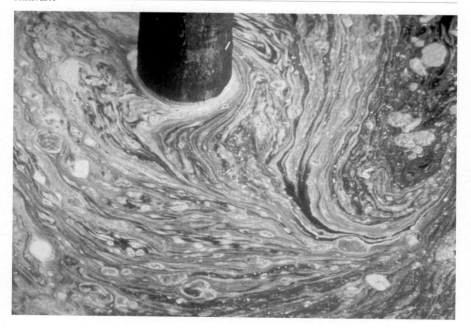

Photo: Mort and Alese Pechter–

Non-point source pollution. Pictured: Run-off in Puerto Rico

year in an area of twelve thousand square kilometers. Lack of oxygen so totally decimated bottom dwelling organisms in the New York Bight (an area stretching sixty-five hundred square kilometers) that recreational divers began noticing abnormally high mortality amongst scallop, lobster, and quahog populations. Scuba divers and fishers also reported that some species of fish which normally inhabit sand and rock bottoms had migrated to surface waters in a desperate attempt to survive.

Although a long standing affliction, sewage only captures world headlines when it encroaches upon human comfort and safety. Such was the response in the 1980s when the public was alarmed and outraged because beaches in Long island and New Jersey were suddenly inundated with dead and diseased fish and bottlenose dolphins, dirty syringes, and other hazardous medical waste, resulting in beach closings and quarantines.

INDUSTRIAL WASTE

Industrial pollution, which can emanate from both point sources and non-point sources (such as fallout of gaseous pollutants spewn into the atmosphere from industrial smokestacks), is one of the most destructive and widespread environmental impacts of the twentieth century.

Industrial waste, which includes poisonous metals such as mercury, tin, and lead, is primarily from factories, mines, and to a lesser extent, boats. These "heavy metals" and "high tech" chemical compounds (synthetic organics) not only kill some species outright, but also contaminate survivors and the food chain. This contamination causes chronic problems like disease, deformities, and lesions; it also weakens the immune systems of marine inhabitants and impairs their ability to resist their natural enemies.

Scientific evidence indicates chlorine as a common link in many of the world's most notorious environmental toxins; dioxin, DDT, Agent Orange, PCBs (poisonous chemicals used to cool machinery), and the ozone-destroying chloroflourocarbons (CFCs) are all based on chlorine.

Organochlorines, synthetic chlorine substances, are used in many industries because they are highly resistant to natural breakdown. But this same stability allows them to remain in the environment for decades or centuries. As a result, the entire planet has become blanketed with chlorine-based poisons. From the North Pole to the deep oceans, organochlorines can be found in the water, air, and food chain. Because these synthetic poisons concentrate in

Industrial pollution

Photo: Michigan Sea Grant–

–Photo: NOAA

Sea birds soaked with oil cannot maintain body heat or fly to forage for food.
Pictured: Oil soaked bird after the Exxon Valdez spill.

fat, they multiply as they move up the food chain. Beluga whales in the St. Lawrence River, which are at the top of the food chain, have accumulated such high levels of PCBs that dead whales are classified as legally hazardous waste.

PCBs have been banned in many countries, but still are widely in evidence. Scientists have found that animals with PCBs in their bodies sicken more easily. In 1988, for example, over eighteen thousand common seals in the North Sea died of disease. Autopsies revealed that most of these seals had PCBs in their bodies.

Experts predict that industrial pollution will continue to be a problem in the future, and in fact, will probably increase as industrial development increases and coastal populations grow. The Environmental Protection Agency (EPA) regulates the amount of pollutants that may be discharged into the marine environment through the National Pollutant Discharge Elimination System. This system requires polluters to obtain permits which fix limitations on the quantities of pollutants that may be discharged, prohibits the discharge of others, and requires the polluter to file periodic reports on their operations, among other requirements.

OIL AND HAZARDOUS MATERIALS SPILLAGE

Petroleum compounds and their by-products are also sources of marine pollution. Whereas some oil naturally seeps from the sea bed, most is from human activities. Experts estimate that each year between three and six million metric tons of oil are discharged into the ocean from atmospheric, land, and sea-based sources.

When oil spills into the sea, it forms a large pool, called a *slick*, on the surface of the water. The slick can drift to shore, foul beaches, kill wildlife, and alter marine ecosystems. Unfortunately, public awareness of the damage wrought by oil pollution is often limited to dramatic and large scale oil disasters, such as the *Exxon Valdez.*

In March 1989, when the huge oil tanker *Exxon Valdez* ran aground in Prince William Sound, Alaska, 250,000 barrels (45 million liters) of oil poured into the sea. The death toll from this one spill alone included 150 rare bald eagles, 1,000 sea otters and hundreds of thousands of fish, seals, and shellfish. Over 100,000 sea birds died of hypothermia and related causes as their feathers became clogged with oil and tar balls and lost their insulative ability. The attendant damage to other sea life and the vital fisheries in the area is still being studied, but juvenile populations are clearly depressed and fewer healthy offspring are surviving despite massive cleanup efforts. Furthermore, cleanups often cause more damage than the oil itself.

In 1991, during the Gulf War, up to two million barrels of crude oil were deliberately pumped into the Persian Gulf from wells in Kuwait, killing acres of sea bottom and marine life. Although these major environmental catastrophes grab the headlines and raise world consciousness, there are actually thousands of less spectacular tanker accidents and other incidents every year. Additionally, a large percentage of oil enters the ocean from oil rigs, cars, machinery and dredging operations, as well as from routine boat operations such as cleaning of boat ballasts and bilges.

The number of vessels legitimately moving petroleum and hazardous substances, both tankers and tank barges, and the number of port calls and other movements are large. The latest available figures show that there are more than 25,000 ships of over 1,000 gross registered tons in world's merchant fleet. About 5,500 of these are tankers, including 265 U.S. flag tankers.[1]

A concurrent unresolved environmental problem related to the transport of oil is that of floating tar. Floating tar generally comes from tankers flushing out their ballast tanks before entering port to take on new cargoes of petroleum. This poses a serious hazard for marine wildlife; young sea turtles seem especially vulnerable and are often found stranded with their mouths gummed with tarballs.

In addition to oil transported by vessels, the U.S. contains a large number of oil refineries which are located along coasts, salt marshes, or estuaries. Accidental oil loss or seepage from refineries contributes 100,000 tons of oil annually to ocean pollution. There are numerous other issues associated with offshore oil and gas operations, including drilling wastes, atmospheric pollution, and canals which are dug through the wetlands for oil pipelines. An additional problem of oil production and storage is also plaguing several European countries-the disposal of obsolete oil rigs. As recently as July, 1995, a team of environmental commandos from the Greenpeace Organization barricaded themselves onto the *Brent Spar*, a 460-foot steel and concrete oil storage platform off the coast of Amsterdam to stop Royal Dutch/Shell from scuttling the rig into the North Atlantic. Scuttling the *Brent Spar* would have set a precedent for more than 400 oil rigs that may become obsolete as North Sea oil and gas run out. Luckily, the dramatic escapade helped win public opinion and five European governments forced Shell to seek a more ecologically responsible (but more costly) alternative.

AGRICULTURAL WASTE

Agricultural pollution usually reaches the ocean from a non-point pollution source, such as unregulated runoff carrying agricultural pesticides, herbicides, and fertilizers. This pollution also contains nutrients and fecal coliform bacteria, bacteria found in human and animal wastes which can cause serious illness. Pesticides and herbicides are not readily biodegradable, and therefore remain concentrated as they pass through the marine food chain. The nitrates from fertilizers foster blooms of algae and decrease the amount of dissolved oxygen in the water. It has been reported that 150 pounds of nitrogen per acre emanate from agricultural fields every year.

Nutrient runoff from agriculture is particularly devastating to coral reefs. If too much nitrogen falls upon reef-building corals, multi-cellular seaweeds start to grow on them. The symbiotic algae that live within the coral polyps are then shaded out, and the coral dies. As they are beaten by the waves, the exposed limestone skeletons are broken up, and the sand thus formed smothers yet more areas of reef.

Although modern agriculture can be extremely harmful to the environment, it should be noted that old-fashioned farming methods also caused pollution. Nitrogen from manure applied to crops could "run off" and pollute adjacent aquatic environments just as easily as modern fertilizers sometimes do. What counts most of all is the amount of fertilizer which is applied and the timing of application. Agricultural science, used responsibly, can result in success stories. For example, research is currently underway in several countries, such as England, to devise methods of fertilizing crops for high yields, without run-off, and to find ways of controlling pests without using vast quantities of noxious pesticides.

RADIOACTIVE WASTE

Another dramatic concern is that of ocean-dumped radioactive waste from nuclear power stations. Even though the large concentration of ocean-dumped radioactive material is within the coastal zone, currents and winds can carry this material into the open ocean. Particularly high concentrations of nuclear waste are found in the Arctic and North Atlantic Oceans, the North and Barents Seas, and many Arctic and Russian rivers. Even though individual concentrations of toxic waste may not be immediately lethal, they are biocumulative, that is, they remain highly dangerous for thousands of years, and not only pose a threat to lower marine organisms, but can contaminate the entire food chain and eventually cause cancer and death in humans. Radioactive pollution is one of the ironies of modern technology: by trying to improve upon nature, we may in fact destroy it.

Nuclear testing became a new and highly controversial activity after World War II. Although the acute impacts of radiation were frighteningly apparent, the long-term effects were unknown. Missile testing (both above and underground), bombing practice, and the construction of nuclear bases all had detrimental effects on the marine environment. In 1979, on the atoll of Moruroa, in French Polynesia, a trial explosion took place, using a bomb of Caesium-134, which has a half-life of two years. The explosion immediately dislodged a million cubic meters of coral and rock from the side of the atoll, causing underwater avalanches and tidal waves. In 1990, scientists still found high levels of radiation contaminating the adjacent environment.

Although the Ocean Dumping Act barred high-level radioactive waste dumping, between 1946 and 1970, the U.S. was allowed to dump over 110,000 packages of plutonium and cesium into its waters, most of it near densely populated urban areas. These dumping sites included Massachusetts Bay near Boston, the Farallones Islands near San Francisco, as well as two sites within tree miles of

Newark, New Jersey. In 1990, NOAA reported that 25% of the 47,500 barrels of atomic waste which had been dumped in the gulf of the Farallones National Marine Sanctuary, had ruptured, and posed an unmitigated threat to many of the valuable commercial fish stock, sea birds, and marine mammals in the area. Once considered the richest marine habitat in the West, since 1991, it has cost the U.S. government almost a billion dollars to conduct a study to assess the extent of the damage from radioactive material.

The Partial Test Ban of 1963 stopped atmospheric testing in the U.S. and England and was followed by a similar ban in France in 1974. Unfortunately, even today, some underground testing still takes place in other parts of the world.

In 1983, a global agreement (an annex to the London Dumping Convention of 1975) was reached which banned dumping of radioactive waste at sea. However, hazardous radioactive waste is still transported by ship. The potential for a large-scale, virtually irreversible disaster, is thus unallayed. The former Soviet Union, for example, recently admitted to having dumped seventeen nuclear reactors and thousands of tons of liquid and solid nuclear waste into the Kara and Barents Seas throughout the "Cold War," even after the global ban on radioactive dumping was enacted. Russia still has nuclear submarines afloat, which although officially decommissioned, are housing reactors which may not be dismantled for many years.

Nor has the problem of eventual leakage of previously discarded containers of nuclear waste been adequately addressed. Lake Karachy in central Russia, long used as a dump for a nuclear weapons plant, now holds radioactive material equal to twenty-four times the amount of the total fallout from the Chernobyl disaster, and scientists fear a high risk of a catastrophic washout. Nor do the mistakes of the past always shape the wisdom of the present. Much to the chagrin of environmentalists and world leaders, in the spring of 1995, France announced that it plans to resume nuclear testing in the far Pacific.

ATMOSPHERIC POLLUTION

"Damage already done to the ozone layer will be with us, our children, and our grandchildren throughout the twenty-first century."

– Former British Prime Minister Margaret Thatcher

Air pollution is an insidious problem. Every day, enormous amounts of combustion exhaust fumes from petroleum which is used for many forms of transportation and power production are spewn into the atmosphere. The EPA reports that U.S. industries pump at least 2.4 billion pounds of chemicals into the air every year. Large quantities of these airborne toxins wind up in the marine environment. Some marine chemists estimate that sixty to eighty percent of the contaminants in highly urbanized areas, such as the New York Bight, comes from atmospheric inputs. Airborne and land-based pollutants travel significant distances, and as they make their way up the food chain, they contaminate a wide spectrum of marine organisms and fish, and ultimately, humans.

"ACID RAIN"

" Acid rain spares nothing. What has taken humankind decades to build and nature millennia to evolve is being impoverished and destroyed in a matter of a few years–a mere blink in geologic time." — **Don Hinrichsen**

Few environmental issues have generated so much controversy as acid rain (also called acid deposition). ACID RAIN is produced from the burning of fossil fuels by power plants, industry, and motor vehicles, whose smokestacks and tail pipes release millions

Acid rain is created when noxious compounds are created by the combustion of fossil fuels and are spewed into the air, such as from smokestacks, combine with water vapor.

Photo: EPA –

of tons of sulfur and nitrogen oxides into the atmosphere. Every year, approximately 160 million tons of these exhaust gases are released into the global atmosphere as a result of human activities. When these nitrogen and sulfur exhaust gases mix with air-borne water vapor, they are converted into sulfuric and nitric acids and eventually return to earth in some form of precipitation. Natural rainfall has a pH level between 5.0 and 5.6; rain below a pH of 5.0 is classified as "acid." Depending on emission rates and weather patterns, this solution can be carried hundreds or thousands of miles in the atmosphere before dropping to the ground as deadly acid rain. Acid rain directly damages whatever it lands on over time, including fields, homes, and people.

The first victims of acid rain are often lakes, ponds, and streams, especially those whose bedrock and soils have a low supply of chemicals to neutralize the acidic deposition. Thousands of lakes in highly industrialized countries are so acidic that they have become totally devoid of fish. In the United States, one out of every five lakes is seriously acidified. In Sweden, 40,000 of its 90,000 lakes have been contaminated by acid rain.

Acid rain finds its way into the marine environment both by direct atmospheric pollution and by migration of freshwater through estuaries into the ocean. Controlling nitrogen and sulfur oxides is particularly difficult because their sources are so varied and dispersed. Further, because acid rain does not respect geographic borders, assessing blame and retribution has created heated political disputes. The countries which generate the highest amounts of fossil fuel emissions are the U.S., Russia, Poland, East and West Germany, the U.K., Canada, and China, but the resulting acid rain pollutes neighboring countries and eventually the entire biosphere. Acid rain causes billions of dollars of damage. In the United States alone, damage to water, soil, food crops, buildings, and human health from acid rain is costing over ten billion dollars a year.

"THE GREENHOUSE EFFECT"

"Humanity is conducting an unintended, uncontrolled, globally pervasive experiment whose ultimate consequence could be second only to nuclear war."

– World Conference on Changing Atmosphere, Toronto, 1988

The GREENHOUSE EFFECT is the term used for global warming that results when accumulations of carbon dioxide, normally processed by plants into oxygen, and other gases, such as methane and nitrous oxide, trap solar infrared light (heat) in the atmosphere. It is called the "Greenhouse Effect" because, like a pane of glass in a greenhouse, these gases let in visible light from the sun, but prevent some of the resulting infrared radiation from escaping and reradiate it back to the earth's surface.

This is negatively affecting the entire planet's ecological balance by changing climactic patterns. According to many scientists, greenhouse gases have already committed the earth to an average warming of between 1° and 4°F above that of 150 years ago, before the industrial era began. By and large, the earth's inhabitants have not felt this increase yet because the oceans have absorbed the heat. Other experts speculate that erratic cooling trends may be camouflaging the full extent of the global warming. Scientists fear that if we continue to consume coal and oil at the present rate, a global temperature rise of 3° to 10°F within fifty years is quite conceivable.

There is much debate amongst scientists over whether the claims of catastrophic damage from global warming have been exaggerated. Some scientists also contend that carbon dioxide is less pivotal to global warming than water vapor. An indisputable fact, however, is that carbon dioxide causes more water to evaporate from land and sea. The water vapor in turn traps more heat, thus amplifying the effect of the carbon dioxide. Climatologists also know that once carbon dioxide is introduced into the atmosphere, it remains there for centuries, so whatever climactic effect it may have will be irreversible for generations. This phenomenon is exacerbated by the documented accompanying depletion in the ozone layer, another controversial environmental issue to consider when discussing pollution.

DEPLETION OF THE OZONE LAYER

The ozonosphere, a stratum of a naturally-occurring oxygen compound located about twenty miles above the earth, acts as an ultraviolet radiation filter and atmospheric insulator. This essential boundary layer is adversely affected by petrocarbon exhaust gases as well as by chloroflourocarbons (CFCs)–chemical compounds used in everything from aerosol hair sprays to refrigerants, fire extinguishers, and solvents–and by natural disasters like forest fires and volcanic eruptions.

"Holes" in this layer allow more damaging solar radiation to reach earth, further contributing to overall temperature increases, as well as triggering an increase in skin cancers and other radiation

effects in humans, animals, and plants. Other types of genetic muta-
tion and defects are also attributable to increases in background
radiation and climactic change.

These abstract environmental phenomena thus have real consequences for all earth's inhabitants. Perhaps the most serious consequence of such damage, from the standpoint of the marine environment, is the projected rise in sea level due to thermal expansion of
the oceans and the melting of the polar ice caps. One of the most
compelling and often asked questions about the marine environment is: "is the world sea already rising and if so, by how much?"
Researchers at Scripps Oceanographic Institution have been
attempting to answer that question by analyzing statistics in several
different regions of the oceans, where sea level data recorded by
tide gauges appears to be relatively homogenous. The conclusions
so far show that the average sea level in the world ocean may
indeed be rising by a few millimeters a year.

A by-product of this process is an expected massive sediment
and nutrient runoff. The results may include damage to and loss of
low-lying coastal regions such as the eastern U.S., coral reefs, mangroves, salt marshes, and dependent marine species which would
unlikely be able to adapt to the concurrent rise in sea level, the
reduction in salinity, and the increase in turbidity.

In effect, the consequences from the greenhouse effect are not
understood, as there are many feedback systems, such as increased
cloud cover, which may result in net increases or decreases in temperature. Further, anthropogenic emissions may very well lead to a
global cooling. The main point is that these emissions are changing
the atmospheric composition of our planet at alarming rates, and
we as yet do not know what the consequences will be. Many scientists are more concerned over the cumulative indirect effects of this
type of industrial pollution and concomitant world-wide deforestation and resulting erosion than over any isolated accidents and
incursions of whatever magnitude. The bottom line is that causing
significant global climactic modifications will have serious effects
on us all.

Marine Debris

Marine debris is not new. As far back as the era of the ancient
mariners, the sea has always been used as a garbage dump. What is
particularly troublesome today is the *massive amount* of garbage
being generated and the *type* of garbage. The world's population
and industrial growth of the last hundred years has produced a
quantity of marine debris unrivaled in history. Exactly how much

debris has actually accumulated in our oceans and waterways? Unfortunately we do not know, as there are no current quantitative studies. In 1975, however, the National Academy of Sciences estimated that the ocean was being bombarded with 14 billion pounds of litter a year, or almost three times the weight of the entire annual catch of fish and shellfish in the United States. A more recent study estimated that the world merchant fleet alone dumps more than five and a half million metal, glass and plastic containers into the ocean every day.

It has also been reported that at least 160 species of marine vertebrates and two species of invertebrates ingest marine debris, and up to a million sea creatures are killed every year by debris thrown into the sea. As the world's population continues to grow, more and more debris will be generated, and that debris will kill growing numbers of marine wildlife.

Historically, trash that ended up in the ocean was made of paper and cloth which decayed, or metal and glass which sank and disappeared from view, making it easy for offenders to ignore the problem. The NIMBY (Not In My Back Yard) principle gave many people a false sense of security. As a result of modern technology, however, our oceans and waterways have been inundated with a material which gives marine debris not only high visibility, but also longevity, *plastic*.

The very characteristics which make plastic a household and industrial wonder make it a disaster in the ocean. Plastic is strong, it is lightweight, it does not rot, it does not rust, it is resistant to ozone, and it is not biodegradable. It can literally last up to hundreds of years. Plastic debris is in evidence everywhere. In coastal cleanups in the Gulf Of Mexico, for example, volunteers routinely collect plastic debris which originates in twenty-eight different countries, including locations as distant as Japan, Bulgaria, and Antarctica.

The Center for Marine Conservation estimates that at least half of all marine debris today consists of manufactured plastic items and plastic resin pellets (plastic resin pellets are the raw form of plastic, typically in the shape of spherules or beads, that have been synthesized from petrochemicals). The plastic items most commonly found are fishing gear, packaging materials, plastic bags and bottles, balloons, and syringes.

Widespread use of plastics has only been in effect for 40 years. In just the last ten years, the use of plastics in packaging has more than doubled. In 1975, nearly 5.6 billion pounds of plastics were used in packaging. In 1987 this figure soared to 15.2 billion pounds. In this relatively short space of time, some of the most remote and

Photo: Jill Townsend/CMC –

The Center for Marine Conservation estimates that up to 80% of all marine debris today is made of plastic and other synthetic materials.

pristine oases on earth, such as Antarctica, have be-come visibly despoiled with debris.

Plastic and other types of marine debris originate from many sources. In addition to merchant ships, Navy vessels, which typically hold thousands of military personnel, generate about 1,000 pounds of trash for every 5,000 people. Commercial fishing boats abandon or lose over 100,000 tons of "ghost" fishing gear (netting, traps, monofilament line, buoys, and cyalume sticks) every year. Recreational vessels are also responsible for marine debris. The U.S. Coast Guard estimates that fifty-two percent of the trash dumped into U.S. waters is from recreational vessels. Even though waste disposal from offshore oil operations is strictly regulated, marine debris associated with petroleum activities is routinely sighted.

PROBLEMS WITH PLASTIC DEBRIS

I. ENTANGLEMENT

Plastic lines, nets, traps, and packaging materials entangle and kill untold numbers of fish, marine mammals, and sea birds. Fishing gear made of synthetic materials, which is purposely or unintentionally abandoned, may drift near the surface or remain on the ocean floor. This is referred to as GHOST FISHING. Ghost fishing gear can continue to entrap marine life indefinitely. Large plastic items such as diapers and bags can smother corals. Monofilament line can entangle and tear soft sponges and sea fans.

II. INGESTION

Plastic items are also ingested by marine creatures. Whales and turtles in particular may swallow plastic bags or balloons, mistaking them for jellyfish. Ingestion of plastic bags is particularly dangerous for sea turtles, since the plastic may get stuck in the turtle's throat and asphyxiate it, or block the turtle's digestive tract and cause death from starvation. Researchers have also found that approximately fifty of the world's two hundred and eighty species of sea birds are known to eat plastic and feed plastic to their young, especially Styrofoam pellets, which they mistake for fish eggs and plankton.

For recreational user groups, garbage in and around the water poses immediate and serious problems. Broken glass, rusty metal, and contaminated waste can cause infection and

Synthetic fishing gear kills over a million marine creatures a year.

Photo: Stephen Frink–

disease if it penetrates into the body. Divers and swimmers who become entangled in monofilament fishing line can drown. Marine debris can damage and incapacitate recreational boats and other vessels by fouling propeller shafts and blocking water intake ports for propulsion or cooling systems. Vessels have even been sunk as a result of collisions with especially large debris items.

HUMAN POPULATION GROWTH

"The real issue is how many footprints will fit on the earth?"

– U.S. Senator Alan Simpson

Population growth and the world capacity for food production are increasing constraints to the marine environment. Demographic analysis shows that it took hundreds of thousands of years for the human race to reach a population level of 10 million, only 10,000 years ago. This number grew to 100 million people about 2,000 years ago and to 2.5 billion by 1950. Within less than the span of a single lifetime the population has more than doubled to 5.5 billion in 1993, and that number is expected to double again by the middle of the twenty-first century. In the last decade, food production from both land and sea declined relative to world population growth.

This accelerated population growth resulted from rapidly lowered death rates combined with sustained high birth rates. This was possible because of increases in food production and distribution, improvements in medical technology and public health, and gains in education and standards of living in many developing nations.

The relationships among human population, economic development, and natural environment are complex and not fully understood. Nonetheless, there is no doubt that the threat to the biosphere is linked to population size and resource use. Increasing greenhouse gas emissions, ozone depletion, acid rain, loss of biodiversity, habitat destruction, excessive mining of minerals, fossil fuel consumption, and shortages of water, food, and fuel indicate how natural ecosystems are being pushed ever closer to their limits.

HABITAT ALTERATION AND DESTRUCTION

"An unnatural decoupling of humans and nature is one of the unfortunate results of high population density and urbanization."   *– Eugene Odum*

Major Pollutants Affecting U.S. Coastal Waters

POLLUTANT	SOURCE	EFFECTS
Nutrients, including nitrogen compounds	Fertilizers, sewage, acid rain from motor vehicles and power plants	Creates algae blooms, destroys marine life
Chlorinated hydrocarbons; pesticides, DDT, PCBs	Agricultural runoff, industrial waste	Contaminates and harms fish and shellfish
Petroleum hydrocarbons	Oil spills, industrial discharge, urban runoff	Kills or harms marine life, damages ecosystems
Heavy metals; arsenic, cadmium, copper, lead, zinc, mercury	Industrial waste, mining	Contaminates and harms fish
Soil and other particulate matter	Soil erosion from construction and farming; dredging	Smothers shellfish beds, blocks light needed by marine plants
Plastics	Ship dumping, household waste, litter	Strangles, mutilates wildlife, damages natural habitats

Source: *Newsweek*, August 1, 1988.

Photo: Doris Alcorn/CMC –
A monk seal with its mouth sealed shut by a plastic packing ring dies of starvation.

In conjunction with population expansion, one of the greatest environmental problems we face today is habitat alteration and destruction. Although some natural events (erosion, subsidence, and storms) alter habitats and ecosystems, human activities are the major culprit in habitat degradation. Deforestation, coastal construction, mining, land fill and dredging operations, the building of roads and drainage ditches, year-round grazing of ranges, and the cultivation of easily eroded lands are destroying valuable terrestrial, marine, and coastal ecosystems.

In addition to eliminating habitats, these activities may result in excessive runoff of rainwater, which can cause floods. They may also lead to drought, which occurs when too little water is stored underground. Moreover, runoff strips soil from the land and deposits it in reservoirs, ship channels, and other bodies of water. These silt-laden bodies must then be dredged or abandoned.

Destruction of habitats, direct and indirect, occurs daily. In many parts of the world, coastal, marine and freshwater habitats are regularly dynamited and bulldozed to create housing, industrial facilities, airports, ports, and resorts for burgeoning populations. In Southeast Asia, live coral is dug up, pulverized, and used as construction material. Other poor developing nations build urban and industrial complexes literally on top of living reefs. Indirect damage

–Photo: Stephen Frink

A sea fan infested with algae from pollutants in the water.

Dredging operations cause silt and toxins to run off into the ocean.

Photo: EPA–

to aquatic habitats is perpetuated by the effects of construction–uprooting and runoff of sediments and toxic chemicals, siltation of reefs which depend on clear water and sunlight, and alterations in water salinity, flow, and temperature. Coral reefs in Southern Florida are declining from being inundated with excessively salty water as a consequence of the diversion of fresh water from the Everglades. Instead of flowing into Florida Bay, the water is being used for agriculture and to supply the growing population along the Florida coast.

Wetlands have been systematicly destroyed, with the blessing of the federal government, since the days of colonial America. In 1764, the Virginia Assembly chartered the Dismal Swamp Company (of which George Washington was a member) to drain 40,000 acres of the Great Dismal Swamp for logging. In the mid-1980s, the U.S. Congress allocated 65 million acres of wetlands for the states to sell to increase their revenues. Between 1940 and 1960, the U.S. Department of Agriculture subsidized the drainage of another 60 million acres of wetlands for agriculture.

Habitat destruction often goes hand in hand with tourism, and specifically the tourist market which brings in the most revenue. In 1994, for example, the governor of the state of Quintana Roo in Cozumel, Mexico, approved a proposal to build a cruise ship dock to cover Paradise Reef, a mile and a half of living reef and one of Cozumel's most popular and prolific dive sites. Even though local dive operators and other groups lobbied against the construction project (which was originally canceled after the President of Mexico was sent a petition with 5,000 signatures), government officials reversed their position and gave approval to build the pier because they believed tourism was more important than protecting the ecology.

The total impact of habitat destruction is sometimes underestimated. By zeroing in on the most obvious and immediate problems, we sometimes overlook the subtle but equally or more insidious ones. For example, one of the worst effects of logging, dredging, dumping, and mining operations is not the felling of trees or removal of minerals and coral, but the construction of dirt roads, camps, piers, and other infrastructures that support these enterprises (especially where the activity takes place on steep slopes), not to mention the energy used up and the waste generated by the construction crews and equipment.

Solutions to habitat degradation are by no means simple. Competition between humans and the environment is an extremely difficult dilemma, with economic and political, as well as scientific, ramifications. In the U.S., for example, over half the population lives within fifty miles of the sea. Roughly two thirds of the world's people live along coastlines and rivers draining into coastal waters. And it is the coastal areas within 200 miles of land that contain the most productive ecosystems. These areas account for more than half the ocean's biological productivity and supply nearly all the world's catch of fish.

The destruction of wetland habitats is particularly harmful to the biosphere. Wetlands (bays, estuaries, mangroves, and marshes) which are spawning grounds for seventy-five percent of commercial seafood species, are routinely drained and bulldozed to make way for farmlands and industrial, municipal, and recreational development. Because of human activities, by the mid 1970s, the U.S. had lost 50% of all its wetlands, and it continues to lose another 371,000 acres each year.[2] Only nine percent of California's original 3.5 million acres of coastal wetlands remains today. Wetland destruction is financially as well as ecologically costly. The federal government estimates that ongoing wetland losses cost the nation's fisheries more than $200 million annually in reduced catches.

Pollution is also a widespread cause of habitat degradation. Coastal waters are heavily polluted throughout heavy urban areas from New York and New Jersey to California. Beyond the United States, contamination of coastal waters is highly evident in Japan, Brazil, the Baltic Sea, the North Sea, the Mediterranean, and elsewhere. Runoff from dredging, filling, and building associated with coastal development operations can affect water temperatures and salinity, and inundate the marine environment with organic and inorganic toxins and sediment, at the same time smothering the seagrasses which would normally filter out these harmful materials. Sediment promotes the growth algae blooms, including toxic "red tides." Dredged material may also contain PCBs, heavy metals, oil and grease, pesticides and pathogens which contaminate the food chain.

Human-induced alterations in habitats caused by the construction of dams and sea walls can also be highly destructive. Species such as salmon, for example, need cool river water, but dams allow the water to become warmer. Other, less useful species then dominate the water and prey upon small migrating salmon that must pass downstream through the reservoir. Modern engineering works, such as dams, are also responsible for a reduction of nutrient and sediment supplies which rivers carry to estuarine and coastal ecosystems, resulting in declining fish populations.

Diseased coral in the Florida Keys as a result of pollution.

Photo: Stephen Frink–

DAMAGE FROM COLLISIONS

Marine habitats are damaged and destroyed by physical impacts. In particular, coral reefs are degraded by boat groundings and improper anchoring of boats. Large cruise ships head the list of offenders in this category, but commercial and recreational vessels of all sizes routinely contribute to the problem as well. As recently as August, 1994, a 170-foot research vessel owned by the University of Miami ran aground at Looe Key National Marine Sanctuary in Florida, reducing some of the coral reefs in the area to rubble. Scientists fear that the damaged corals may take years to regrow, if they regrow at all.

Anchors dropped directly on coral can annihilate innumerable living marine resources. As the anchor chain drags across a coral reef and is later retrieved, dozens of feet of coral can be slashed and crushed, destroying the coral outright, or leaving it defenseless against marauding bacteria and algae. Dragging anchor chains also stirs up clouds of silt which smother living coral polyps. (See Chapter 15 for correct anchoring techniques.)

On a much lesser scale, incidental damage to reefs can be caused by unwary or poorly trained divers and snorkelers. In a study done recently in Looe Key, Florida, it was observed that 26% of the divers and 61% of the snorkelers touched coral at least once in a 30 minute period. Five percent of the divers touched coral 20 times or more in 30 minutes.

*Valuable wetlands being
drained and bulldozed
for agriculture.*

–Photo: Raymond Gehrman

The ongoing groundings of commercial and recreational vessels in the Florida Keys, events which were damaging miles of valuable living coral reef, finally prompted legislators to designate the area a National Marine Sanctuary in 1992. However, successfully protecting an environment, even one which is under formal jurisdiction, requires ongoing funding, effective management, monitoring and research, and cooperation among resource managers and user groups.

TOURISM AND RECREATION

Tourism, one of the largest and fastest growing industries in the world today, is a double-edged sword. In some small countries like Belize, for example, tourists outnumber the local population. On the one hand, tourism provides jobs and financial support in developing nations. Many small islands, such as the Turks and Caicos, derive the majority of their gross national product from tourism. On the downside, tourism puts pressure upon natural habitats and may infringe upon local culture. Mass tourism began in the 1950s with the advent of widely available and affordable transportation. Many places that were once serene wildlife habitats, priceless historic monuments, and fragile ecosystems have become bustling holiday resorts, overrun by millions of travelers whose main concern is their own immediate pleasure.

Tropical marine and coastal settings, with balmy weather, picturesque scenery and facilities for water sports are prime tourist attractions. In Florida, reef tourism brings in approximately $1.6 billion a year, with over 2 million people a year visiting John Pennecamp Coral Reef State Park and Key Largo National Marine Sanctuary. Unfortunately, many of these tropical havens are extremely vulnerable to the effects of tourism and recreational activities. Bright lights, crowds, and collisions with boats, or their anchors and propellers, can harm and scare marine creatures. The intrusion of loud noises, such as from overhead aircraft, boats, water skis, and jet skis, has been shown to interfere with creatures who rely on acoustic sounds as their primary means of communication. Loud noise can also cause seabirds to abandon their eggs or chicks and cause marine mammals and fish to deviate from their normal behavior.

Traditionally, many marine areas have allowed tourists to feed fish and other wildlife. In fact, tour guides and divemasters are often tipped extra by their paying customers for luring fish into public view with Cheez Whiz, bread, eggs, and other artificial delicacies. This "food" disrupts the natural food chain and fish eating habits and can be detrimental to the fish's digestive system. Cutting up and feeding fish "natural" food items, such as sea urchins, is also a habit which should be discouraged because it removes integral links in the reef ecosystem. Feeding fish may also provoke aggressive behavior. There have been reported incidents of divers injured by fish that are conditioned to bite an open hand underwater, whether or not it contains food.

Because divers interface directly with the underwater environment, they are often blamed unfairly for environmental damage. Other than the very limited aforementioned study at Looe Key, we have almost no data to determine to what extent skin and scuba divers actually contribute negative effects.

We do know, however, that dive travel is increasing and that most divers travel to tropical reefs. According to *Skin Diver Magazine*'s Diver Survey, the average diver takes 3.7 trips and stays eight days per trip. The survey also reports that in the last three years, 66% of *Skin Diver Magazine* subscribers traveled outside the continental U.S. on diving trips. Seventy-one percent of dive travelers went to the Caribbean, 51.7% to Mexico, 30% to the Bahamas, and 27% to Micronesia, the Red Sea, and the Great Barrier reef of Australia. Of those who dived in the U.S., 64.7% traveled to Florida. These numbers verify that divers are flocking to areas which are vulnerable to human impacts.

In addition to divers and snorkelers, participants in other recreations such as fishing, boating, water skiing, and jet skiing, frighten marine creatures from their habitats and alter nesting, feeding, and mating patterns. Large numbers of tourists can also result in additional marine debris, sewage pollution, and the removal of living marine resources for souvenirs. However, there is no data which can quantify the amount of damage caused by people vacationing and participating in recreational activities in these environments. Until research can show otherwise, therefore, we must assume that the recreation industry is not the source of major alterations and destruction in the marine environment. Nevertheless, the conservation ethic must be strongly promoted among recreational users. Proper education, training, and orientation at resort areas help prevent accidental damage.

DECLINE OF BIODIVERSITY

The earth contains between three and forty million species, but only 1.4 million of these species have been classified. We are presently on a course in which we are losing nature much faster than we can learn about it. The loss of biodiversity is a crucial concern in marine conservation and is a common theme throughout this book. The term BIODIVERSITY (biological diversity) refers to the diversity of life, and is often divided into three distinct categories:

1. genetic biodiversity (diversity within a species);
2. species biodiversity (diversity among species); and
3. ecosystem biodiversity (diversity among ecosystems).

All three areas of biodiversity are critical. One of the most immediate questions in environmental conservation today is, "How can we best resolve the complex problem of declining species populations and biodiversity before supplies become exhausted?" Scientists estimate that 36,500 species of plants and animals are becoming extinct every year, mostly because of human activities. If terrestrial as well as aquatic habitats continue to be altered and

Photo: Reef Relief–

Loud noises disturb marine wildlife and disrupt their activities.
Pictured: A waterskier disturbing a mangrove.

destroyed at the current rate, at least 500,000 and perhaps 1 million species will become extinct over the next twenty years.[3]

Until recently, however, most environmental attention has focused on the loss of biodiversity in terrestrial ecosystems, especially tropical rain forests. But scientists, economists and legislators, as well as industrial, commercial, and recreational entities, and concerned citizens, are beginning to realize that the threats to the living resources of our world's marine and coastal ecosystems are equally significant, and that loss of aquatic life can depreciate the entire biosphere. Scientists fear that the alarming decline in populations of fish, for example, may set off a chain of ecological disasters affecting not only the fish, but also the animals and humans that feed on fish, and the species on which the fish prey.

Additionally, as scientists continue to validate the ecological principle that earth is a biosphere, collaborative efforts are now underway to gain greater insight into the relationship between the biological, chemical, and physical nature of marine ecosystems. Scientists are starting to take a holistic approach to the problems of the marine environment, considering both the effects of natural disturbances and human interventions on marine communities and the effects of the changes in species populations on the entire ecosystem. As opposed to the traditional single-species focus, now ecosystem-wide surveys are being conducted, such as one sponsored jointly by NOAA and the EPA. These surveys are providing important quantitative data on the key ecosystem components, including phytoplankton, zooplankton, nutrients, and hydrography.

The protection and preservation of biodiversity in habitats, from rain forests to coral reefs, is a global and compelling challenge. Whereas we still do not have a solution that is universal, at least there is increasing agreement that the *problem is universal* and that something must be done soon. If a solution is at all possible, therefore, it will have to be predicated on an unprecedented alliance among citizens, industry, and government.

LIVE ROCK HARVESTING AND FISH COLLECTING

Since prehistoric times, shells and coral have been harvested from the sea for many purposes: decoration, jewelry, tools, even currency. The fascination with these items continues to grow. In coral reef communities throughout the world, thousands of tons of corals and shells, as well as fish for aquariums, are taken from the ocean by amateur and commercial collectors, resulting in a lucrative billion dollar a year trade.

The large aquarium trade is one of the markets that demands tropical marine animals and living reef rocks. Several hundred fresh water and marine species are currently collected, and most of the marine species are taken from reefs. The U.S. is the largest importer of tropical fish in the world, and it obtains about eighty percent of its supplies from the Philippines.

Since the aquarium trade does not pose an immediate danger of extinction to fish or invertebrates, it is not formally regulated. However, fish collecting should only be done by individuals trained in ecologically sound procedures. Poisons, such as sodium cyanide, used in the Philippines to extract fish from their habitat, should be condemned. This toxic chemical damages the liver, kidneys and reproductive organs of the fish (which die within weeks), and it indiscriminately kills corals and other reef inhabitants.

Fish collection may also exact a high death toll due to poor handling and transport. Amateurs who attempt to remove fish from the ocean can injure or kill them. Certain species, for example, must be brought to the surface extremely slowly, or their gas bladders will burst from the change in pressure. Some species should not be removed under any circumstances because they cannot survive outside the reef environment. Half of all butterfly fish, for example, die within two months after capture.

The removal of live coral for aquariums and jewelry is a widespread problem. About 250,000 pieces of live coral were imported into the U.S. in 1991 for the aquarium industry. Most commercially valuable precious coral, which is used for jewelry and statues, comes from the Mediterranean or from deep seamounts in the Pacific.

In many of these countries, the coral grounds are over-exploited, and restrictions on the taking of rare and endangered species are poorly enforced or non-existent. Historically, coral was harvested selectively by divers who would choose from coral colonies according to size and quality. Modern coral harvesting is much more destructive because it often involves non-selective dredging devices such as the "Italian Bar" and the "St. Andrew's Cross," which strip large tracts of the seabed along with its inhabitants.

The countries dealing most heavily in ornamental shells are Indonesia, the Philippines, India, Mexico, Haiti, and Kenya. Often countries which ban the removal of shells and coral from its own reefs, such as the U.S., allow these items to be imported from the Philippines for resale to local collectors. Mother-of-pearl, the thick iridescent layers found inside the shells of several mollusk species, has been prized by collectors for centuries. The best mother-of-pearl comes from reefs in the Indian and Pacific Oceans, and the main suppliers are Indonesia, the Philippines, Australia, the Solomon Islands, New Caledonia, and Papua New Guinea.

Although pressure from coral, shell, and aquarium fish collection are stressful to marine habitats, responsible regulation of these industries can decrease the damage as well as bolster the economy and provide jobs for local artisans in developing countries. In the Philippines, training courses have been set up to teach local collectors alternative and less invasive methods of catching aquarium fish. In some nations, shell collectors are beginning to work with governments to become more selective in what they remove and sell, avoiding endangered species. The CITES (Convention on International Trade in Endangered Species of Wild Fauna and Flora) is observed in over a hundred countries. This law classifies species into two categories, Appendix I of the CITES deals with species which are most threatened and prohibited for trade, and Appendix II deals with species which are allowed to be traded, but are regulated and require documentation.

Another case study of conservation efforts can be seen in Fiji, where the Fisheries Division has issued a set of guidelines for the exploitation of reefs for use other than food harvesting. The guidelines require an exploratory survey before any new coral reef area is harvested. However, smuggling and long-standing exploitation practices still undermine these and other conservation efforts.

DECLINE AND DEPLETION OF FISH POPULATIONS

Ninety-six million metric tons of fishes, shellfish, and algae are taken out of the world's oceans every year by commercial and sport fishers. One hundred million metric tons is the limit of maximum sustainable yield advised by the United Nations Food and Agriculture Organization (FAO). SUSTAINABLE YIELD refers to the optimum annual catch that can be derived indefinitely from harvested species, without causing a stock failure. In addition to the fact that we are almost beyond that limit, almost thirty percent of the catch is never even used for human consumption. An enormous amount of fish is wasted because of poor fishing methods or because it is used in products other than food.

Of the nearly 20,000 known species of fish, about 9,000 are currently harvested, but only 22 species are regularly caught in significant quantities. Just six groups, herrings, cods, jacks, redfishes, mackerels, and tunas, account for nearly two thirds of the total annual catch. Studies by the National Fish and Wildlife Service (NFWS) indicate that 85% of the species currently fished in American waters are now overexploited. At least 14 species of oceangoing fish, including Atlantic salmon, yellowtail flounder, grouper, Spanish mackerel, bluefin tuna, swordfish, and Pacific perch, have been so seriously depleted that it could take them twenty years to recover, even if all fishing were to stop tomorrow.

Dwindling fish supplies present immediate and long-term problems having scientific, political, and economic repercussions. Over half the population of developing nations depends on fish as its primary source of dietary protein. In highly industrialized countries, as well as in developing nations, fishing contributes significantly to economic stability. In the U.S. alone, where one third of the population fishes recreationally, it has been estimated that the indirect and direct impact of just recreational fisheries (not including commercial fishing) adds up to about $50 billion a year and generates 600,000 full-time jobs.

Part of the problem can be attributed to natural causes such as oceanic and weather fluctuations like El Nino-warmed waters which result in reduced food supplies. But experts feel that the most detrimental toll is imposed by human activities–man-made pollution, coastal construction, and excessive and abusive fishing practices.

Many researchers feel that the primary reason that fish populations are declining is that too many people are taking too many fish too quickly. Unlike terrestrial wildlife preserves and parks, living marine resources have historically been free to the public, with little

Bycatch is a serious problem. Pictured: A live loggerhead turtle captured by accident in a fishing net.

–Photo: Mike Weber/CMC

or no restrictions over those who might abuse and exploit these resources. Many countries of the world, particularly those desperate for food and economic survival, employ wanton and wasteful fishing practices, such as ghost fishing, dynamiting coral reefs, and the use of poisons to extract fish from their habitats. These methods cause chronic damage to entire marine ecosystems.

In industrialized countries, sophisticated fishing devices, such as computers, radar, electronic depth finders, spotter planes, and helicopters, allow commercial fishing boats to locate large schools of fish with speed and accuracy. Another modern invention, immense gill nets (some are 40 miles long and 300 to 500 feet deep) sweep everything in their path, including turtles, dolphins, sharks, and sailfish.

TRAWLS, nets with a wide mouth tapering to a small, pointed end, are towed behind a vessel at any depth in the water column. Gillnets have been outlawed in many countries of the world because of their huge bycatch, in which species other than the target species, and juveniles, are incidentally trapped and wasted. The magnitude of the bycatch problem can be seen in shrimp trawling in the U.S. Gulf of Mexico shrimp fishery, where trawl mesh is so small that for every pound of shrimp taken, an average of ten pounds of bycatch is caught as well.

It is also important to understand that biotic impoverishment goes beyond the loss of individual species. Fish are a keystone, or critical link, in the food chain of a marine ecosystem. Excessive removals of fish can have dramatic effects which reach up the food chain to marine mammals and cascade down to plankton. So, although there are still habitats that contain few or no endangered fish species, they may contain so few representatives of each species present that the functioning of an entire ecosystem can be impaired.

There is yet another problem confronting fish populations, one which is not as well understood, but is equally serious, the alteration of the gene pool in fish species. By selectively fishing out the larger, more desirable members of certain species, we may create inferior breeds, dominated by younger, smaller fish which mature and reproduce earlier and with fewer offspring, and have a shorter life span. We may also unwittingly replace commercially prized species with commercially useless species possessing superior survival adaptability.

The decline and depletion of fish populations is a dramatic example of what the American ecologist Garrett Hardin referred to as THE TRAGEDY OF THE COMMONS, the irresponsible plundering of a public resource by individuals interested only in short-term financial profit.

NOTES

CHAPTER

THE U.S. GOVERNMENT
IN CONSERVATION

"We can search for and find solutions that will help generations yet to come....Doubt is a luxury the world can no longer afford."   **– President Bill Clinton**

In response to the growing threats to the global environment, state and federal government agencies in the United States and elsewhere have taken on increased responsibility for conservation. This includes the management of marine fisheries, the establishment of marine sanctuaries, the protection of endangered species, and efforts to reduce pollution and to protect natural habitats.

President Clinton and Vice President Gore are undoubtedly the most committed environmental leaders that America has had for the last twenty years. However, since Bill Clinton assumed the presidency, no major environmental bills have been passed by the U.S. Congress. In any administration, however, environmental legislation is always opposed by high-paid professional lobbyists and lawyers who represent special interest groups, such as developers, large trawling factories, and logging and mining companies. The big businesses that profit from pollution and habitat destruction have considerable resources and continually exert pressure on the House and Senate members.

MARINE CONSERVATION LEGISLATION

" Next to personal example, law is the most powerful teaching tool in society."

– J. William Futrell, Pres. Environmental Law Institute

Although most people have only become familiar with environmental laws fairly recently, there is evidence of environmental legislation which was passed as far back as 2000 BC in Babylonia (pro-

135

hibiting the adulteration of grain). In 1273, England enacted laws limiting the burning of coal, and in the 1700s, the U.S. adopted smoke-abatement ordinances. In the 1940s, the increase in high-tech chemical wastes invoked the fear that modern industry was dangerous to human health as well as to the environment. Rachel Carson's exposé, *Silent Spring,* which was published in 1962, invoked a loud public outcry against DDT and other toxins. But broad pollution regulations in the U.S. did not really begin until the 1970s–the era in which the Love Canal disaster[1] sparked public outrage about statutes such as the National Environmental Policy Act (NEPA), Clean Water Act (CWA), and the Clean Air Act (CAA).

Over the last decade, environmental laws have increased dramatically in number and complexity. Understanding the labyrinthine body of environmental laws which exists today is not an easy task, even for legal experts. The following are helpful guidelines for understanding U.S. environmental laws:

1. environmental laws are enacted by Congress and the state legislatures;
2. environmental laws are usually designed to address one major subject (i.e., the Resource Conservation and Recovery Act (RCRA) deals with hazardous wastes, the Clean Water Act deals with water pollution, etc.);
3. all environmental laws have underlying presumptions about how they will operate; they are based upon either a standard of environmental quality or upon technological achievabililty;
4. in many instances, the essence of environmental law is not in the written statute, but in the regulations. Environmental laws are not self-regulatory. The U.S. government has delegated responsibility to federal agencies, particularly the EPA, to develop specific requirements of its environmental statutes. Environmental regulations are usually promulgated by technical people within the regulatory agency; and
5. almost all federal environmental statutes have provision for citizens to sue for enforcement if the government does not enforce the law.

It is important to keep in mind that as well-intentioned as our environmental laws are, monitoring and enforcing them are difficult and have met with varying degrees of success.

The following two chapters will present an overview of the federal and international efforts for marine conservation which have been put into operation during the last few decades.

POLLUTION CONTROL

Among international, federal, and state conservation laws, one of the most ambitious is MARPOL ANNEX V. MARPOL Annex V addresses the marine debris problem and was signed in 1987 by the United States and thirty-eight other countries (now forty-one signatory countries to date). This law, under the enforcement of the U.S. Coast Guard, bans the dumping of plastics into the ocean or navigable waterways of the United States and limits the over-board disposal of other garbage by all ships of signatory countries. All ships, boats, platforms, marinas, and docks must comply with this law. All marinas and docks are required to provide on-shore trash containers. Vessels twenty-six feet in length and longer must have at least one MARPOL Annex V placard prominently posted, and vessels forty feet or longer must also have a written waste management plan on board.

Although MARPOL Annex V is the mandate which most widely addresses the marine debris problem, several other U.S. laws have also been enacted to prevent wildlife entanglement in plastic debris, such as the Marine Mammal Protection Act, the Endangered Species Act, and the Migratory Bird Treaty Act. The Fishery Conservation and Management Act prohibits the disposal of nets into U.S. waters, and a new international agreement may expand the outlawing of drift net fishing to all other countries of the world as well.

The National Oceanic and Atmospheric Administration (NOAA), as directed by Congress, also plays a role in dealing with marine debris. The National Ocean Pollution Planning Act of 1978, called for NOAA to establish a comprehensive, coordinated, and effective federal program for ocean pollution research, development, and monitoring. NOAA, therefore, in consultation with other agencies, prepares a five-year Federal Plan for the National Marine Pollution Program every three years.

Other major U.S. marine conservation laws, organizations, and agencies include the NOAA National Marine Sanctuary Program, Coastal Zone Management Act, Coastal Barriers Resources Act, Clean Water Act, Ocean Dumping Act, Marine Plastics Pollution Research Control Act, Oil Pollution Act, Endangered Species Act, Marine Mammal Protection Act, and the Marine Entanglement Research Program. (See Appendix I.) Unfortunately, monitoring and enforcing these laws is extremely difficult, and in many instances, unsuccessful.

CONTROLLING OIL SPILLS

In response to the oil spillage problem, in 1990, the U.S. Congress passed the Oil Pollution Act, a law which requires tanker and tank barge operators to phase in the use of double hulls on their vessels. Additionally, the oil and gas industry has set up a response organization for major spills, the Marine Spill Response Corporation (SUPER FUND). Due to these newer and more stringent environmental laws, in September 1994, the Exxon corporation was ordered to pay an unprecedented $5 billion in punitive damages to commercial fishermen, property owners, and Alaskan natives harmed in the 1989 *Exxon Valdez* oil spill.

NATIONAL MARINE SANCTUARY PROGRAM

In response to the growing threats to endangered marine habitats, the National Oceanic and Atmospheric Administration (NOAA) created the National Marine Sanctuary Program (NMSP) in 1972. Since 1972, the Secretary of Commerce has designated fourteen sites as National Marine Sanctuaries, ranging in size from less than one nautical mile to several thousand nautical miles.

The objectives of the National Marine Sanctuary Program are:

1. to protect marine resources through conservation and management;
2. to promote marine research and monitoring;
3. to promote public awareness and wise use of marine environments; and
4. to facilitate multiple uses compatible with resources protection.

The NMSP encompasses an enormous diversity of marine life in over 14,000 square miles of ocean. According to Edward Ueber, manager of the Gulf of Farallones NMS, the national marine sanctuary program protects ninety-five percent of the marine habitat in the United States. Each sanctuary is unique and valuable–from the Florida Keys NMS, which is the third largest barrier reef in the world, to the Flower Garden Banks, which is the only known oceanic brine seep in the continental shelf waters of the Atlantic, to the ironclad Civil War warship, *U.S.S. Monitor*, to Fagatele Bay, a tropical reef surrounding a volcanic crater.

In addition to the existing national marine sanctuaries, NOAA continually researches and recommends to the U.S. Congress new sites to be designated as sanctuaries. To create a sanctuary, NOAA selects a site from a SEL (Site Evaluation List). Each site on the SEL must possess qualities of special national significance based on conservation, recreational, educational, ecological, historic, research, and aesthetic values.

In 1994, Congress increased funding for the NMSP to $9 million. In 1995, funding increased to $12 million. The National Ocean Service supplemented that appropriation with over a half million dollars aimed at better resource protection and management of the Tijuana River, San Francisco Bay, and Monterey Bay watersheds. Unfortunately, the total sum is still meager compared to funding allocated for national parks, and the NMSP budget must be shared by all the marine sanctuaries facilities and staff.

Each national marine sanctuary conducts numerous projects. One such project is a sanctuary user survey at Stellwagen Bank NMS. Some sanctuaries have initiated a joint enterprise with the Coast Guard on oil spill contingency plans. The Monterey Bay NMS is developing a Water Quality Protection Program. In addition to individual sanctuary initiatives, the NMSP is in the process of developing a broad strategic plan to guide marine resource protection into the year 2000. This plan represents the work of sanctuary field and headquarters staff, as well as many partners within government and in the private sector.

One of the most important functions of the national marine sanctuaries is resource management. Because sanctuaries are visited by millions of people each year, sanctuary managers must develop and follow a plan to prevent abuse and degradation of the sanctuary's resources, while fostering research projects, educational programs, and responsible recreational and commercial activities. It is imperative that divers become familiar with these regulations and programs (which may vary from sanctuary to sanctuary) and support them.

Since its inception in 1972, the NMSP continues to serve as a model for marine parks and sanctuaries in many other countries of the world, where governments are seeking to limit commercial abuse and uncontrolled exploitation. Divers, fishers, boaters, and travelers can research these issues globally by finding out about programs in effect in their own countries or travel destinations. Public input and support are critical elements for the success of a sanctuary, and all visitors as well as local residents should become involved as much as possible. For example, in the Florida Keys, the NMSP's largest public involvement effort to date, the organization has accomplished an important milestone with the publication of the Draft Florida Keys Management Plan. The draft management plan reflects years of coordinated involvement with commercial and recreational fishermen, charter boat operators, divers, environmentalists, government officials, citizens, and business people united to bring the best possible resource management to the Florida Keys.

NATIONAL MARINE SANCTUARIES

Channel Islands NMS
Located twenty nautical miles off the coast of Santa Barbara, California, this sanctuary contains 1,252 square nautical miles of off-shore, near-shore, and intertidal habitats, including giant kelp forests. Endemic wildlife includes sea birds, blue whales, humpback whales, seals, and sea lions.
113 Harbor Way
Santa Barbara, CA 93109
(805) 966-7107
Manager: LCDR John Miller

Cordell Bank NMS/Gulf of the Farallones NMS
North of the Farallones, this NMS consists of 397 square nautical miles of ocean waters, surrounding a submerged mountain top, along the edge of the California Continental Shelf. This sanctuary is home to subtidal and oceanic creatures such as fur seals, sperm whales, and sea turtles.
Fort Mason, Building #201
San Francisco, CA 94123
(415) 556-3509
Manager: Edward Ueber

Fagatele Bay NMS
Located in American Samoa, this sanctuary consists of .25 square nautical miles of a coral reef ecosystem which surrounds an eroded volcanic crater.
PO Box 4318
Pago, Pago, American Samoa 96799
(684) 633-7354
Coordinator: Nancy Daschbach

Florida Keys NMS
Extending over 2,800 square nautical miles, from Biscayne Bay to the Dry Tortugas, the Florida Keys NMS is the third largest barrier reef in the world. It supports subtropical ecosystems with an enormous variety of plant and animal life, and it is home to numerous shipwrecks.
Main House, 5550 Overseas Highway

Marathon, FL 33050
(305) 743-2437
Manager: Billy Causey

Flower Garden Banks NMS
This sanctuary, encompassing approximately 100 square nautical miles of ocean south of the Texas/Louisiana border, protects the most northern and ecologically diverse coral reef system in the continental U.S.
1716 Briarcrest Dr., Suite 702
Bryan, TX 77802
(409) 847-9296
Manager: Steve Gittings

Gray's Reef NMS
Located off the coast of Georgia, this sanctuary encompasses seventeen square nautical miles of submerged limestone reef. Gray's Reef NMS protects numerous species of marine wildlife and the only known calving ground for the endangered right whale.
30 Ocean Science Circle
Savannah, GA 31411
(912) 598-2345
Manager: Reed Bohne

Hawaiian Islands Humpback Whale NMS
This sanctuary encompasses approximately 500 square nautical miles of ocean around Maui, Lanai, Molokai, and the waters adjacent to the Kilauea National Wildlife Refuge on the island of Kauai. It protects the breeding and calving grounds for the humpback whale and also provides habitat for other species of whales, dolphins, sea turtles, and fish.
300 Ala Moana Blvd., Room 5350
Honolulu, HI 96850
(808) 541-3184
On Site Liaison: Janice Sesing

Key Largo NMS
This sanctuary, extending 100 square nautical miles in the upper Florida Keys, is home to an abundance of coral reef communities and historic artifacts.
PO Box 1083
Key Largo, FL 33037
(305) 451-1644
Manager: LTCD Paul Moen

Looe Key NMS

Located in the lower Florida Keys, this sanctuary encompasses approximately 5.3 square nautical miles of rich and diverse coral reefs. Looe Key Sanctuary was named for the *H.M.S. Looe,* a British warship that ran aground on the reef in 1744.

Route 1, Box 782
Big Pine Key, FL 33043
(305) 872-4039
Manager: George Schmahl

Monitor NMS

The iron-clad Civil War warship was the first national marine sanctuary in the NMSP. Only 1 square nautical mile in size, the *U.S.S. Monitor* NMS was granted federal protection in 1972 because of its rich historic value.

NOAA
Building 1519
Fort Eustis, VA 23604-5544
(804) 878-2973
Manager: Dr. John Broadwater

Monterey Bay NMS

Encompassing 4,024 nautical square miles along the California coastline, this is the largest of the national marine sanctuaries to date. This sanctuary is home to kelp forests and submarine canyons and contains seals, sea lions, sea birds, blue whales, humpback whales, and sei whales.

299 Foam Street, Suite D
Monterey, CA 93940
(408) 647-4201
Manager: LCDR Terry Jackson

Olympic Coast NMS

Located along the northern Pacific coast of Washington state, this sanctuary is home to a variety of salmon, shellfish, halibut, and cod. It also protects numerous wildlife endangered and threatened species, including bald eagles, peregrine falcons, brown pelicans, sea otters, several species of whales, and porpoises.

138 West First Street
Port Angeles, WA 98362
(206) 457-6622
Manager: Todd Jacobs

Stellwagen Bank NMS

This sanctuary, encompassing 842 square nautical miles between Provincetown, Cape Cod, and Cape Ann, Massachusetts, is home to whales, fish, shellfish, harbor seals, gray seals, sea turtles, and numerous species of sea birds, as well as many shipwrecks.
14 Union Street
Plymouth, MA 02360
(508) 747-1691
Manager: Brad Barr

PROPOSED SANCTUARIES

Norfolk Canyon

Sixty nautical miles off the mouth of the Chesapeake Bay, this site includes a 35 mile long submarine canyon, lush marine biodiversity, and a valuable fishery.
1305 East-West Highway, 12th FL
Silver Springs, MD 20910
(301) 713-3132
Program Specialist: Cheryl Graham

Northwest Straits

North of Seattle, this site is home to numerous species of sea birds, including bald eagles, auklets, and oyster catchers. Migratory fish in the area include salmon, trout, anchovy, pollock, and halibut. The Northwest Straits are also home to dolphins, seals, and pods of orcas.
7600 Sand Point Way, NE
Seattle, WA 98115
(206) 526-4293
On-site Liaison: Linda Maxson

Thunder Bay

Located in Michigan, Thunder Bay may become the first freshwater national marine sanctuary. This area is ecologically as well as historically important because it contains sculpted limestone bedrock, submerged terraces, and scarps, as well as over 100 identified shipwrecks.
Dept. of Parks and Recreation
Natural Resources Bldg., #141
Michigan State University
East Lansing, MI 48823
(517) 336-3142
On-site Liaison: Michele Richhart

U.S. FISHERIES

With the growth of modern fishing fleets in the second half of the twentieth century, it became increasingly apparent that fish stocks would have to be protected. THE MAGNUSON FISHERIES CONSERVATION AND MANAGEMENT ACT 0F 1972 designated eight U.S. regional fishery management councils which set commercial and recreational fishing regulations and management policies. (See Appendix I.) A FISHERY is the combination of fish and fishers in a region, fishing for similar or the same species with similar or the same type of gear.

The Magnuson Act created federal authority over the U.S. domestic fisheries for the first time in our nation's history and changed the way our waters were fished. Most significantly, the Magnuson Act virtually eliminated competition from foreign fishing fleets by giving Americans exclusive fishing rights within a 200-mile conservation and management zone around the U.S. shoreline, known as the Exclusive Economic Zone (EEZ).

THE EIGHT REGIONAL FISHERY MANAGEMENT COUNCILS

New England Fishery Management Council
5 Broadway (Route One)
Saugus, MA 10906
(617) 231-0422
Executive Director: Douglas Marshall
Jurisdiction: Connecticut, Maine, Massachusetts, New Hampshire, Rhode Island

Mid-Atlantic Fishery Management Council
Federal Building, Room 2115
300 South New Street
Dover, DE
(302) 674-2331
Executive Director: John Bryson
Jurisdiction: Delaware, Maryland, New Jersey, New York, Pennsylvania, Virginia

South Atlantic Fishery Management Council
One South Park Circle, Suite 306
Charleston, SC 29407-4699
(803) 571-4366
Executive Director: Robert Mahood
Jurisdiction: East Florida, Georgia, North Carolina, South Carolina

Caribbean Fishery Management Council
Suite 1108, Banco de Ponce Building
Hato Rey, PR 00918-2577
Executive Director: Miguel Rolon
Jurisdiction: Puerto Rico

Gulf of Mexico Fishery Management Council
Lincoln Center, Suite 331
5401 West Kennedy Blvd.
Tampa, FL 33609
(813) 228-2815
Executive Director: Wayne Swingle
Jurisdiction: Alabama, Louisiana, Mississippi, Texas, West Florida

Pacific Fishery Management Council
Metro Center, Suite 420
2000 Southwest First Avenue
Portland, OR 97201
(503) 326-6352
Executive Director: Lawrence Six
Jurisdiction: California, Oregon, Washington

Western Pacific Fishery Management Council
1164 Bishop St., Suite 1405
Honolulu, HI 96813
(808) 541-1974
Executive Director: Kitty Simonds
Jurisdiction: Hawaii

North Pacific Fishery Management Council
605 West 4th Avenue, Room 306
Anchorage, AK 99510
(907) 271-2809
Executive Director: Clarence Pautzke
Jurisdiction: Alaska

The major objective of the Magnuson Act was to ensure that fish can be caught while at the same time preventing overfishing. The eight regional Fishery Management Councils, each with similar and different needs, are charged with this task. These Councils are comprised of federal, state, private-sector individuals, and other experts appointed by the Secretary of Commerce from recommendations by coastal state governors. Additional expertise is provided by Advisory Panels and Scientific Committees. An important factor which helps balance the decision-making process is the citizen input which the Councils allow.

Current methods by which fisheries are attempting to deal with the decline of fish populations include:

1. imposing catch limits;
2. gear restrictions;
3. size limits;
4. seasonal fishery closings;
5. quotas; and
6. limiting the number of fishing licenses.

Unfortunately, many of these traditional management actions have proven inadequate or impractical, especially in areas of great fishing pressure. Limiting the number of fishery licenses issued, allocating quotas (total annual quotas can be divided into individual transferable or non-transferable percentage quotas allotted to fishermen), and bag limits can be expensive to monitor and difficult to enforce. They also require timely and accurate data and precise knowledge about the various species and the fishery.

There are critical gaps in fishery catch statistics, both in terms of the amount of information collected and the adequacy of the collection systems. These gaps deny fishery managers essential information on the current levels of commercial and recreational harvest, and on fish discarded as well as landed. These research and information shortfalls are largely the result of chronic underfunding. Poor funding is also responsible for insufficient studies of habitat and ecosystems. As suggested by the laws of ecology, to understand fish, one must study their interdependent relationships to their environment. This means studying prey/predator interactions and the effects of selectively and intensively removing certain species from an ecosystem. Research is needed to assess the effects of altering the physical and chemical environment on fish behavior, growth, feeding, and reproduction.

Selective fishing gears (e.g., mesh size, hook size, etc.), limits on the number of sets of gear allowed per boat, and limited entry, along with most approaches, may reduce fishing effort, but still tend to select against certain species and the larger individuals within the species. Closed seasons and temporarily closed areas may not be effective because fishing efforts can be increased at other times or in other localities. Finally, bag limits and size limits can be ineffective due to unintentional release mortality. Some reef fishes die when caught in deep water because of injuries associated with depth and temperature change. Even when handled carefully, a relatively large percentage of fish will die because of the way they are hooked or because they are damaged in the net or on the boat deck.

NO-FISHING RESERVES

Some fisheries are considering instituting no-fishing zones, a concept which has been established successfully in other parts of the world including the Great Barrier Reef in Australia, Bermuda, New Zealand, South Africa, the Philippines, and other countries. Fishing Reserves are areas which are protected so that fish can produce gametes which result in larvae and juvenile fish, which can then extend out or migrate from the sanctuary to provide the basis for recruitment in other areas being heavily fished.

AMENDMENTS TO THE MAGNUSON ACT

Developing and enforcing management policies in fisheries remain ongoing problems, due to lack of funds needed to gather scientific information (i.e., data on fish population size, sustainable yield, and bycatch levels) and complicated by political pressures, small versus large scale commercial enterprises, and recreational and scientific constituencies. The National Marine Fisheries Service estimates that although commercial and recreational fisheries contribute more than $30 billion annually to the U.S. gross domestic product, these valuable public resources are largely being squandered.

New fisheries management actions directed to the long-term economic and ecological sustainability of fish species are under consideration, but have not yet been ratified by Congress. The Marine Fish Conservation Network, in conjunction with the National Audubon Society, Greenpeace, CMC, and other environmental groups, is currently lobbying to re-authorize and strengthen the Magnuson Fishery Conservation and Management Act. Their goals are to:

1. increase funding for monitoring and enforcement activities;
2. enact a universal licensing program to give managers necessary information on who's fishing, when, where, and how, and what they are catching;

3. enact a comprehensive at-sea observer program to monitor commercial fisheries to provide unbiased and detailed information on fishing activities as they occur; and
4. focus on measures that are enforceable at the dock or the point of sale, rather than measures that must be enforced offshore.

In 1994, to meet these objectives, Congressman Wayne Gilchrest (R-Maryland) proposed an amendment to the Magnuson Act, HR 4404. Congressman Gilchrest's amendment bill contained many critical provisions under five main categories:

1. eliminating overfishing and rebuilding depleted fish populations;
2. reducing incidence of bycatch to near zero, by requiring new selective gear, allowing the National Marine Fisheries Service to provide both incentives for fisheries which use gear that avoids bycatch (including TEDs-Turtle Excluder Devices and FEDs-Fish Excluder Devices) and "disincentives" (penalties) for fisheries that continue to take bycatch;
3. reforming the Regional Fishery Management Councils to seat more members who do not have financial conflicts of interest, and to give votes to individuals who represent the interests of non-user and conservation groups;
4. protecting marine fish habitats; and
5. providing adequate funding for fisheries research, management, and enforcement.

"Aquaculture" and "Mariculture"

One solution to dwindling fish supplies has been practiced in China for over four thousand years–AQUACULTURE and MARICULTURE. Aquaculture is the farming of fish and other marine organisms in ponds, estuaries, bays, or open ocean close to shore. Mariculture refers specifically to the farming of fish in the sea. Aquaculture has increased steadily in the United States during the past twenty years, with rapid advances in salmon production. Aquaculture now produces more than a third of a million tons of salmon yearly. Aquaculture is also big business for several other species, such as catfish, and there are commercial farms in over thirty states. But aquaculture has its opponents–environmentalists who are concerned about the escape of wastes, chemicals, antibiotics, and alien species into the water, and local fishermen who fear that aquaculture infringes on their livelihood.

ARTIFICIAL REEF PROGRAMS

The use of ARTIFICIAL REEFS, ecosystems created around intentionally submerged structures–which can range from abandoned boats, planes, trains, oil rigs, and steel bars to obsolete military craft–is viewed with ambivalence by many scientists. Based on the natural shipwrecks on the ocean floor which have become nuclei of marine life, artificial reefs in tropical sites are placed either directly on a damaged reef, in the hopes of helping it to regain its biomass, or on a seabed in proximity to a reef. The wreck's structural material (usually metal) becomes a substrate for coral larvae to settle on, and the wreck eventually replicates a natural habitat, providing shelter, a source of food, and an area for mating and reproduction for a variety of creatures.

Scientists fear that some of the materials used to create artificial reefs, such as discarded tires and municipal solid waste ash from incinerators, may in fact decompose and contaminate the very reef community they are intended to support. More research is needed before these concerns can be supported or refuted.

Aquaculture, or fish farming.

Photo: Mort and Alese Pechter–

Photo: Mort & Alese Pechter –

An army tank which will be sunk to create an artificial reef.

THE CORAL REEF COALITION

With the support of Vice President Al Gore, author of *Earth In The Balance,* an innovative international initiative began in January 1994. The Coral Reef Coalition, which was held in Washington DC, for the first time brought together scientists, legislators, resource managers and representatives of business, the World Bank, and private user groups from over two dozen nations to seek cooperative solutions for conservation of coral reefs throughout the world. The invited participants (including the author) attended plenary sessions, followed by group discussions, and concluded with group and individual recommendations for Vice President Gore and Timothy Wirth, US Undersecretary for global affairs, continues to reconvene the group on a regular basis.

REAUTHORIZATION OF CONSERVATION LAWS

Appendix 1 of this book lists all the major U.S. environmental laws which affect the conservation of the seas and marine wildlife. In addition to the Magnuson Fishery Conservation and Management Act, three other important statutes are up for reauthorization by Congress in the coming year–the Marine Mammal Protection Act, the Endangered Species Act, and the Clean Water Act.

I. MARINE MAMMAL PROTECTION ACT

This act is designed to protect dolphins, manatees, otters, and other marine mammals. A key issue in the reauthorization is bycatch of marine mammals in fishing nets. However, conservationists and fishing industry representatives reached agreement in June 1993 on proposals that require fishing fleets to reduce significantly their take of marine mammals. The fishing industry also agreed to participate on conservation teams working toward the recovery of declining species.

Artificial reefs teem with marine life. Pictured: Federnmalis wreck in Aruba

Photo: Mort & Alese Pechter –

II. ENDANGERED SPECIES ACT

Although it is most often thought of as a law to protect terrestrial plants and animals, the ESA also provides protection for marine species, a number of which are currently listed as endangered or threatened. Conservationists support a package of strengthening amendments offered by Rep. Gerry Studds (D-MA) and Sens. Max Caucus (D-MT) and John Chafee (R-RI). The two bills (HR 2043 and S 921) would broaden the Act to focus more on ecosystems, rather than single species, and would improve the listing and recovery process.

III. CLEAN WATER ACT

This law regulates water pollution and use of the nation's wetlands, both of which have impact on the health of the nation's coastal waters. Much of the reauthorization debate has centered on Section 404, which regulates the filling of wetlands. Conservationists support HR 350 and S 1195, offered by Rep. Don Edwards (D-CA) and Sen. Barbara Boxer (D-CA), which would expand Section 404 to cover other wetlands activities, including draining, dredging, and flooding. A coalition of conservation and labor groups also supports HR 1720 and S 815. These clean water/jobs bills would help improve water quality in the nation's estuaries by funding improvements to municipal sewage systems and restoration of wetlands.

A heated controversy on Capitol Hill is threatening to overturn the Clean Water Act, currently referred to by environmental groups as the "Dirty Water Bill." H.R. 961, which was introduced in spring of 1995 by Chairman Bud Shuster (R-PA) of the House Committee on Transportation and Infrastructure and passed by the House of Representatives, abandoned the long-standing goals of the Clean Water Act to restore all the nation's waters to swimmable and fishable quality. The new amendments compromise clean water in many ways. They:

–eliminate protection for wetlands by redefining the term "wet lands," and weaken enforcement measures for remaining acres;

–allow point source polluters to decrease their efforts to cut pollution. Allow environmental goals to be discretionary;

–require no treatment for storm water discharges, even where
  pollution from urban storm water exceeds pollutants from factories and sewage treatment plants;
–eliminate the goals for reducing polluted runoff. While grant
  money may increase for state programs, it exculpates states
  from being accountable for those programs to actually reduce
  pollution. The new law only requires states to achieve "reason
  able progress" toward achieving water quality goals in 19
  years. Additionally, that requirement can be waived if
  Congress does not provide 100% of the funding;
–permanently exempts cities under 100,000 people from requirements to reduce storm water pollution;
–do not ensure a citizen's right to sue polluters or the right to
  know what pollution is being discharged into a community's
  water. There is also no provision to phase out the use of highly toxic chemicals, such as dioxin.

(NOTE: as of this book's publication date, the Senate has yet to
approve these amendments to the Clean Water Act. If the bill passes
the Senate, President Clinton has threatened to exert his veto
power).

NOTES

INTERNATIONAL CONSERVATION INITIATIVES

"We cannot change unless we survive, but we will not survive unless we change."

– New York Public Interest Research Group

UNITED NATIONS ENVIRONMENTAL PROGRAM

The United Nations Environmental Program (UNEP) is the premier international organization dedicated to the protection of the environment in general and to the marine environment in particular. UNEP's Regional Seas Program seeks to develop unique regional arrangements to control pollution and manage marine resources. The Regional Seas Program encompasses 10 regional seas with over 120 nations participating. The Regional Seas Program is an action-oriented program encompassing a comprehensive approach to marine and coastal areas and to environmental problems concerning not only the consequences but also the causes of environmental degradation.

The Regional Seas Program has had several impressive successes, such as the Mediterranean Action Plan, the South Pacific Regional Environmental Protection Convention, and the Convention for the Protection and Development of the Wider Caribbean Region (commonly called the Cartegena Convention). These programs control, reduce or prevent marine pollution from vessels, land-based sources, storage of toxic wastes, nuclear testing, seabed exploration, erosion and atmospheric contamination. Another UNEP Program currently underway is the Global Plan of Action for Conservation, Management, and Utilization of Marine Mammals.

LAW OF THE SEA CONVENTION

This is an ambitious treaty that has, to date, been ratified by thirty-five nations, not including the United States. A minimum of sixty signatories is required for the treaty to go into effect. among the broad range of provisions that attempt to set up a system of international marine law are: the regulation of offshore resource development; pollution regulations and liability; scientific research; and marine mammal protection. If adopted, this convention could be the most significant law applying to the world's oceans.

LONDON DUMPING CONVENTION

This convention, forbidding the dumping of certain kinds of waste from ships and aircraft, was implemented in 1975. Signed originally by thirty-three nations, the treaty covers all the world's oceans beyond national jurisdictions, and as of 1988 had nearly sixty signatory parties. The treaty specifically bans the dumping of "black-listed" substances, such as heavy metals, petroleum products, and carcinogens. These substances may by dumped only in "trace" quantities. In addition, a "gray list" of substances including lead, cyanide, and pesticides may be dumped only with specific permits. Since 1983, the LDC has also placed a moratorium on dumping low-level radioactive wastes.

INTERNATIONAL MARITIME ORGANIZATION (IMO)

This organization has developed various treaties to improve marine safety and to control pollution. Among the issues it covers are liability and compensation from oil spills, the charting and removal of offshore oil rigs, the designation of marine areas for special protection, and plastic debris in the water.

INTERNATIONAL WHALING COMMISSION (IWC)

This commission was established by the adoption of the International Convention for the Regulation of Whaling in 1946. The IWC, whose membership now stands at thirty-eight nations, regulates the hunting of whales by establishing catch and size limits and gear specification, and it disseminates information on whale biology. In 1986, the IWC proposed a moratorium on commercial whaling, which was opposed by Japan, Norway, and the former Soviet Union. There is still controversy over "scientific research" and commercial whaling, which are still practiced by several nations, and which the IWC has been unable to stop

UNITED NATIONS FOOD AND AGRICULTURE ORGANIZATION

The fisheries division of the FAO collects information on the entanglement of marine resources in fishing nets and debris, and works with many other UN organizations to gather and coordinate data on marine debris and pollution issues.

ANTARCTIC TREATY PROGRAMS

In 1959, twelve nations adopted the Antarctic Treaty in order to deal with the potential use of the Antarctic continent for military purposes. This treaty defers all claims of sovereignty to lands and waters of Antarctica, prohibits use of the continent for military purposes and nuclear testing, and promotes the freedom of scientific research. Since then, natural resources (i.e., exploration and exploitation of minerals, oil, and gas) have become the dominant international political focus in Antarctica, and amidst much controversy, several new treaties, some adopted by as many as thirty-three nations, have addressed these issues.

INTERNATIONAL UNION FOR THE CONSERVATION OF NATURAL RESOURCES

Another international conservation initiative is IUCN, whose goals are to provide independent international leadership for promoting effective conservation of nature and natural resources and to preserve species, ecosystems, and biodiversity. The IUCN is composed of governments, government agencies, and nongovernmental organizations. IUCN conducts numerous projects related to biological diversity around the world, many of which are related to the marine environment.

THE BRUNDTLAND COMMISSION REPORT

One of the most ambitious attempts to promote conservation on a global scale was a report issued in 1987 by the United Nations World Commission on Environment and Development, under the direction of Dr. Gro Harlem Brundtland, then Premier of Norway.

The study concluded that solutions to the global problems of environmental pollution and natural resource depletion require extreme changes in agricultural, banking, and industrial practices, as well as reduced energy consumption and population growth. It urged the world to move toward SUSTAINABLE DEVELOPMENT, which emphasizes economic growth in an ecologically sound manner, especially in the less-developed countries.

FIRST UN WORLD POPULATION CONFERENCE IN BUCHAREST, 1974

The first of several United Nations environmental conferences took place in Bucharest in 1974, bringing together official government delegations from 136 countries. Growing environmental awareness and developmental pressures were the context for a heated debate. Industrialized nations considered creating measures to control population growth rates the primary objective, while developing nations demanded that economic issues be considered foremost, arguing that "development is the best contraceptive." Nonetheless, the delegates ratified the World Population Plan of Action (WPPA), a broadly drawn set of recommendations and principles.

SECOND UN CONFERENCE ON POPULATION IN MEXICO CITY, 1984

In 1984, 147 countries sent delegates to the second United Nations Conference on Population in Mexico City to revise and extend the WPPA. Public opinion had shifted to favor government population politics, but the issue had become enmeshed in internal U.S. debate over abortion and contraception. Ironically, the positions taken in Bucharest were reversed in Mexico City. While most developing countries acknowledged the need for family planning and population stabilization, the United States–under pressure from anti-abortion activists at home–declared population a "neutral phenomenon" in development.

The U.S. "Mexico City policy" curbed financial aid to family planning groups worldwide. And in 1985, the U.S. cut funding for the United Nations Population Fund. As a result, population control efforts around the world lagged behind growing demand. By 1992, the United Nations Population Fund could only meet a third of its requests for population and family planning assistance.

THE UN EARTH SUMMIT IN RIO DE JANEIRO, 1992

The largest and most successful global summit to address poverty, pollution, and biodiversity was convened in Rio de Janeiro, Brazil for thirteen days in June of 1992 by the United Nations General Assembly. This historic conference was attended by 118 heads of state, including President George Bush, a Global People's Forum representing thousands of nongovernmental groups from all over the world, and more than 9,000 news reporters.

The Earth Summit focused on the imbalance between the "North" and "South," or the developed nations versus the underdeveloped nations. As a result of this conference, an Earth Charter was signed, as were treaties dealing with the preservation of endangered species and habitats and the reduction of greenhouse gases.

The Earth Charter defines principles of conduct to promote a sustainable future. In essence, it is an agreement by which Northern countries can gain access to biological resources, such as chemicals and genes from wild plants in return for transferring money and technical assistance for conservation and sustainable use of wild areas in the South.

Agenda 21, a nonbinding 800-page blueprint to protect the environment and encourage sustainable development, was adopted by a consensus of the 180 nations represented. The agreement, known as the biodiversity treaty, commits these countries to these goals:

1. to draw up national strategies for conservation;
2. to integrate plant and animal preservation into economic planning;
3. to set up protected areas and promote the protection of entire ecosystems as well as individual species;
4. to manage biological resources outside of protected areas;
5. to take steps to restore degraded ecosystems;
6. to take inventory of biological resources and monitor them; and
7. to allow countries that are rich in species but poor in cash to share in the profits from sales of pharmaceutical products or other products derived from their biological resources.

Also adopted was a nonbinding Declaration of Environment and Development which consists of twenty-seven broad principles for preserving biodiversity and giving priority to needs of developing countries. A binding treaty to limit emissions of "greenhouse" gases like carbon dioxide and methane was passed, but the U.S.

insisted that no specific timetables be included. Other industrial countries, however, promised to cut carbon dioxide emissions back to 1990 levels by the year 2000.

Another legally binding treaty was passed, which requires inventories of plants and wildlife, including endangered species. The U.S. was the only participant that refused to sign the biodiversity treaty, on the grounds that it could adversely affect U.S. biotechnology corporations.

Many who attended the Earth Summit came away with a feeling of optimism for future cooperation among nations. To date, however, much of the financial and technical assistance which had been promised has not yet been delivered.

THE INTERNATIONAL SCIENTIFIC CONGRESS IN NEW DELHI, 1993

Representatives of fifty-eight national academies of science from around the world met in New Delhi in 1993 for a "Science Summit" on World Population. The conference grew out of two earlier meetings, one of the Royal Society of London and the U. S. National Academy of Sciences, and the other an international conference organized by the Royal Swedish Academy of Sciences. Statements published by both groups expressed a sense of urgent concern about the expansion of the world's population and concluded that if current predictions of population growth prove accurate and patterns of human activity on the planet remain unchanged, science and technology may not be able to prevent irreversible degradation of the natural environment and continued poverty for much of the world.

The New Delhi conference, organized by fifteen academies, was convened to explore in greater detail the complex and interrelated issues of population growth, resource consumption, socioeconomic development, and environmental protection. It was the first large-scale collaborative activity undertaken by the world's scientific academies. At this conference a statement which reflects continued concern about the intertwined problems of rapid population growth, wasteful resource consumption, environmental degradation, and poverty was signed by the fifty-eight academies. The academies agreed that a prerequisite to dealing successfully with global social, economic, and environmental problems was a stable world population. The goal proposed was zero population growth within the span of the next generation.

The International Conference on Population and Development (ICPD) in Cairo, 1994

Inspired by the unanimous agreement and support of the 58 scientific academies which met in New Delhi in 1993, the United Nations organized the ICPD summit meeting in Cairo, Egypt, in September 1994. The themes of this conference were stabilizing world population and improving social, economic, and personal well-being while preserving fundamental human rights and the ability to live harmoniously in a protected environment. Of the more than 150 delegates attending, the Vatican and several Arab and South American countries registered reservations about the language on sex and abortion. There was tremendous consensus on the other issues. Specific topics addressed at the conference were:

1. the empowerment of women in sexual, social, and economic life, so they can make individual choices about family size;
2. universal access to convenient family planning and health services and a wide variety of safe and affordable contraceptive options;
3. encouragement of voluntary approaches to family planning and elimination of unsafe and coercive practices;
4. clean water, sanitation, broad primary health care, and education;
5. appropriate governmental policies that recognize longer-term environmental responsibilities;
6. more efficient and less environmentally damaging practices in the developed world, through a new ethic that eschews wasteful consumption;
7. pricing, taxing, and regulatory policies that take into account environmental costs, thereby influencing consumption behavior;
8. the industrialized world's assistance to the developing world in combating global and local environmental problems;
9. promotion of the concept of "technology for environment";
10. incorporation by governments of environmental goals in legislation, economic planning, and priority setting, and incentives for organizations and individuals to operate in environmentally benign ways; and
11. collective action by all countries.

After nine days of debate and deliberations, the Cairo convention produced a 113-page declaration, representing a twenty-year "program of action." Under this program, natural and social scientists, engineers, and health professionals will play a part in developing a better understanding of the problems, options, and solutions for this new population policy. These professionals will be specifically charged with helping to deal with:

1. cultural, social, economic, religious, education, and political factors affecting reproductive behavior, family size, and family planning;
2. impediments to human development, especially social inequities, and ethnic, class, and gender biases;
3. global and local environmental change, its causes (social, industrial, demographic, and political) and policies for its mitigation;
4. improving education and human resource development, with special attention to women;
5. family planning programs, new contraceptive options, and primary health care;
6. transitions to less energy- and material-consumptive economies;
7. building indigenous capacity in developing countries in the natural sciences, engineering, medicine, social sciences, management, and interdisciplinary studies;
8. technologies and strategies for sustainable development;
9. networks, treaties, and conventions that protect the global commons; and
10. world-wide exchanges of scientists in education, training, and research.

THE INTERNATIONAL CORAL REEF INITIATIVE

The ICRI workshop convened May 28 to June 2, 1995, in Dumaguete City, Philippines. It was the first major international meeting to bring together leading coral reef scientists, managers, NGOs, international agencies and funders to develop coordinated strategies for coral reef conservation. Participants from thirty-six countries met to review the condition of the world's reefs, consider new scientific findings, assess the effectiveness of current reef conservation and management strategies, and identify priorities for coral reef conservation in the coming decade.

The ICRI hammered out two final documents entitled, *Call to Action* and *Framework for Action*. *Call to Action* describes the significance of coral reef ecosystems and the major threats to the reefs. The *Framework for Action* sets forth the international priorities for reef conservation through improved management, capacity building, research and monitoring. The *Framework for Action* will be used as the foundation document for international program administered or supported by organizations such as the Asian Development Bank, the World Conservation Union, the U.N. Environmental Program, and the World Bank (all of which participated in the Workshop). It is also being used as a starting point for the ICRI Regional workshops which are planned for coming years.

Major topics addressed at the 1995 ICRI include:

1. Dive-Tourism: participants explored the possibility of using dive tourism as an environmentally and economically sustainable alternative to more destructive uses of the reefs, but warned against the false promises of "pseudo-ecotourism," that is not environmentally responsible.

2. The Internet: the group explored the potential for expanded communication of information about coral reefs through data sharing on computer databases and through the establishment of Internet links among those concerned about coral reef conservation.

3. International Year of the Reef: 1996-1997 was officially declared the year of the Reef. This year-long program involves environmental activities and events which will bring together NGOs as well as scientists and government entities. In conjunction with the International Year of the Reef, the Public Awareness and Education Committee was created and met for the first time during the Workshop in order to develop a plan to increase public awareness of the threats to coral reefs.

<u>NOTES</u>

CONSERVATION IN THE DIVE INDUSTRY

"Divers, perhaps more than any other group of people, have the motivation and insights to become the marine environment's most dedicated ambassadors."

– Al Hornsby, Director PADI Project AWARE

Divers can be an excellent resource when it comes to marine conservation. Diving is a recreational activity that is growing rapidly. Currently, there are an estimated 3.5 to 5 million active scuba divers world-wide and at least 10,000 active skin divers, and these numbers increase every year.[1] Because of their large population and because they tend to be well-educated, relatively affluent, and concerned about the environment, divers have the ability to be a powerful constituency group for environmental legislation. Most sectors of the diving industry–the training agencies, equipment manufacturers, retailers, educators, diving book publishers, dive travel agencies, and the diving population at large–are involved in some form of marine conservation.[2]

DIVE TRAINING AGENCIES

Dive training agencies are vital players in marine conservation. In the last few decades, dive training agencies have become more involved not only in environmental education, but in environmental legislative decisions and monitoring programs.

There are well over a dozen dive training agencies in the world, the largest ones being PADI, NAUI, SSI, and YMCA. Many skin and scuba diving courses, particularly entry level training, include discussion of environmental awareness. Instructors attempt to dispel myths and misconceptions, for example that sharks are our enemy, dangerous predators to be feared and killed. Divers are taught that marine creatures are by nature indifferent to or afraid of humans and only attack to defend themselves from perceived danger.

Divers learn early on that "the most dangerous creature in the ocean is a human being."

Divers are taught in the classroom about aquatic environments, and in pool and open water skill sessions, divers are shown techniques which promote minimal impact to the environment, such as buoyancy control, streamlining, correct finning, and breathing techniques which will not cause siltation and disturb delicate habitats. Dive boat etiquette is always stressed, including correct disposal of garbage.

Although a goal of all diving agencies is the preservation and protection of the marine heritage, several agencies have specific marine conservation programs and teaching materials.

PADI

PADI (Professional Association of Diving Instructors), the largest dive training agency in the world (with over 65,000 diving professionals worldwide), oversees the industry's broadest conservation initiative. Under the premise that "one diver can't protect the aquatic environment, but 10 million divers can," PADI created Project AWARE in 1992. The Project AWARE mission statement identifies PADI's goals:

1. to provide PADI members with information, including educational and support materials, to educate recreational divers on the aquatic realm and how to preserve it;
2. to support the efforts of environmental organizations whose missions are aligned with Project AWARE;
3. to support key environmental legislation;
4. to incorporate environmentally sound practices throughout PADI Headquarters internal operations and to encourage the recreational scuba diving industry to do the same; and
5. to provide direct financial support (where possible) to environmental issues and organizations.

True to these objectives, PADI has integrated conservation into its open water teaching materials and developed an entire "Underwater Naturalist" course, a video which teaches divers to perfect their buoyancy control, and other conservation materials. PADI Project AWARE has also teamed up with NOAA for their brochure, "Ten Ways A Diver Can Protect the Underwater Environment," which is distributed at dive resorts and sanctuaries around the world. PADI Project AWARE also works with the Center for Marine Conservation on numerous underwater cleanups around the world, and it awards money and an annual Environmental Awareness Award to worthy recipients. PADI's *Undersea Journal,* a quarterly publication, contains articles which promote environmental awareness and conservation.

PADI also operates through a number of independently-owned international offices. For example, PADI Canada finances artificial reef projects in partnership with the Underwater Society of BC and the Artificial Reef Society of BC, along with help from the parks department. PADI Europe mounted a global effort to protect sea turtles and is sponsoring an Italian research group monitoring marine life in the Mediterranean Sea.

NAUI

In 1992, NAUI signed an ongoing written agreement with NOAA to jointly fund and develop environmental programs and projects. The first project from this alliance was the production of two marine conservation videos, "To Preserve and Protect the Last Frontier" (for divers) and "You Can Make A Difference" (for children). NAUI also offers specialty programs in Underwater Ecology, Conservation Diver, Kelp Ecology, Manatees, Stingrays, and Coral Reef Ecology.

NAUI, a membership organization, also created its own non-profit environmental entity, Environmental Horizons, in 1992, under the direction of the author, Hillary Viders. NAUI's International Underwater Foundation selects individual and group winners for the NAUI annual Environmental Enrichment Award. NAUI's journal of underwater education, *Sources,* a bi-monthly magazine, contains a regular "Environmental Facts" column as well as periodic articles on conservation issues.

YMCA

In the 1970s, the YMCA debuted its Reef Ecology Program, the first ecology program ever for recreational skin and scuba divers. It was developed by Dr. Rena Bonem, a noted marine biologist at Baylor University in Texas. The Reef Ecology Program provides both diving professionals and sport divers with a comprehensive understanding of these fragile ecosystems by means of classroom sessions and five open water dives. In addition to updating all of its curricula every eighteen months, YMCA is now incorporating material on conservation into all of its open water and leadership courses.

In 1994, YMCA introduced an innovative slide and audio presentation entitled "Piggy Divers" which uses cartoons and humor to educate divers on environmentally correct behavior. Topics include correct debris disposal, buoyancy control, equipment configuration, and reef etiquette.

DIVE EQUIPMENT MANUFACTURERS

Many dive equipment manufacturers now use recycled paper and biodegradable materials in their offices and for packaging their products. Dive equipment factories are reducing or eliminating the use of toxic chemicals in production.

In the last few years, some manufacturers have also begun to introduce environment-friendly equipment. DACOR Corporation's "Reef Saver" console, for example, has an innovative design which allows dive hoses and gauges to conform securely across the diver's body, so that they do not hang loose or dangle, thereby preventing collision with sensitive marine organisms and habitats. This design also allows the diver to view his gauges without having to use his hands, which enhances buoyancy control and a streamlined profile.

Minimal impact diving skills include buoyancy control, a streamlined profile, correct weighting, streamlined equipment configuration, and correct finning technique.

Photo: Pete Nawrocky–

−Photo: Pete Nawrocky

Buoyancy control is not just for beginners. It is a skill which divers of all levels should continually practice and fine tune. Pictured: author demonstrating advanced buoyancy control in the BTSI Diamond Reef System during a strong surge condition.

Photo: Pete Nawrocky −

The fin pivot is a good skill for learning buoyancy control.

The "warm water" streamlined buoyancy compensator was created by Seaquest and is now available from several dive equipment manufacturers. This lightweight and compact BC design greatly improves a diver's buoyancy control and trim.

A company from Seattle, Buoyancy Training Systems Inc., has pioneered a system of lightweight and easy-to-assemble diamonds, which are used by many diving instructors to teach buoyancy control, the most important skill to prevent divers from colliding with and damaging coral. The BTSI program uses seven diamonds, which range from fairly narrow to several feet wide, which are positioned to float at different depths underwater. Students must swim through progressively narrower diamonds without his/her body or equipment touching the sides. After students master the basics of buoyancy control, they can swim through the diamonds backwards or in pairs while buddy breathing. The diamonds are part of an entire buoyancy training program. When students complete the program successfully, they are awarded a certificate and customized environmental stamp, a collector's item which is recognized by the National Philatelic Society.

Another new eco-friendly piece of equipment is the Aqua Flare from Princeton Tec. The Aqua Flare, a personal indicator light for night diving, is becoming a popular replacement for the plastic encased cyalume chemical glow sticks which were often thrown overboard after one use.

Some dive equipment manufacturers have parent corporations which fund environmental causes (e.g., Scuba Pro, Inc. is owned by Johnson Corp., which gives generous donations to environmental causes).

DIVING INDUSTRY FINANCIAL SUPPORT

There are dive industry awards, scholarships, and foundation grants designated for conservation efforts. For example, Rodale's *Scuba Diving Magazine* annually awards $10,000 for environmental achievement. Each year, PADI Project AWARE Foundation donates over $100,000 to a variety of environmental protection, education, and research projects. NAUI and PADI/SeaSpace present environmental enrichment awards annually, and the Diving Equipment and Marketing Association (DEMA) has its own Ocean Futures division which raises funds for various environmental projects through an annual silent auction.

In 1995, the Underwater Cleanup and Conservation Monitoring Program began a new precedent – a broad coordinated effort of the diving community, federal agencies (EPA, NOAA, and the US Coast Guard), and the non-governmental environmental sector (Center for Marine Conservation). This annual global shoreline and underwater cleanup program is jointly funded by the PADI Foundation and PADI Project AWARE, DEMA's Ocean Futures, *Skin Diver Magazine*, and the Center For Marine Conservation.

There are also numerous diver-supported environmental organizations (see Appendix 2 for complete list of NGOs). These include the American Oceans Campaign, Reef Relief, Cousteau Society, C.O.R.A.L., etc. See Appendix for complete list. These groups generate financial and public support for environmental causes by supporting conservation events, disseminating literature, and conducting their own fund-raising activities.

DIVE PUBLICATIONS, DIVE CLUBS, AND INDIVIDUAL DIVER ACTIVITIES

All diving publications, particularly Rodale's *Scuba Diving, Discover Diving, Scuba Times Magazine, Ocean Realm,* and *Skin Diver Magazine,* publicize and support issues in aquatic conservation. They are also invaluable in notifying the public of upcoming environmental events in which they can participate. To support conservation on the home front, almost all diving publications now print on recycled and/or chlorine-free paper, and use recycled products in their offices.

There are many hundreds of dive clubs ranging from small local entities to large international groups, such as the British Sub Aqua Club, whose membership numbers in the thousands. Either working alone or with schools, aquariums, and maritime museums, dive clubs conduct and participate in conservation activities. A noteworthy example is GREAT (Gulf Reef Environmental Action Team, a small volunteer group in Texas). GREAT members have been installing mooring buoys in the Flower Garden Bank and Stetson Bank as well as educating the public in the use and importance of these buoys. Other conservation activities in which dive clubs can be involved include:

1. shoreline and underwater cleanups (see Chapter 17);
2. "Adopt A Beach" programs;
3. recycling demonstrations;
4. search and recovery of ghost fishing gear;
5. underwater photography of marine debris and damaged habitats;

A diver recovers lost fishing gear.

Photo: Tom Campbell–

Sea Camp gives youngsters hands-on experience in marine education and conservation.

–Photo: Mort and Alese Pechter

6. rescue of entangled and stranded marine wildlife;
7. leading citizen action at the local, state, and federal levels (see Chapters 20 and 21);
8. working with managers of sanctuaries and protected reserves; and
9. working with scientists to do underwater research and surveys.

One of the most important contributions which divers can make is to educate friends, family, and the general public about conservation. This is particularly important for children, as the knowledge and values which we instill in the younger generation will influence the future of our environment. There are several noteworthy programs organized and conducted by divers specifically for children, such as Sea Camp in Florida and the nation-wide Ocean Pals Poster Contest and Program.

DIVER ROLE MODELS

The diving industry has an array of luminaries who serve as role models and spokespeople for environmental issues. Jacques Cousteau and his son, Jean-Michel, Sea Hunt star Lloyd Bridges, Ted Danson, marine biologists Dr. Eugenie Clark (the "Shark Lady"), Dr. Sylvia Earle (famed aquanaut and deep sea technology pioneer), and Dr. Kathy Sullivan (NASA astronaut and U.S. Chief Scientist), Stan Waterman (five time Emmy award winning cinematographer whose work was seen in *Blue Water White Death, The Deep,* and several National Geographic Explorer films), Marty Snyderman (whose underwater cinematography touched millions of people in the film *Free Willy),* and Robert Ballad (Woods Hole Oceanographic Institute Director who discovered and explored the *Titanic* and other historic shipwrecks) are some of the best known ambassadors for the marine environment. They are celebrities who headline shows and make news headlines.

In addition to these diving super stars, there are countless other divers who have also made important inroads in conservation–diving educators, underwater photographers, cinematographers, and naturalists, such as Norine Rouse, John Halas, Al Giddings, Howard Hall, Mort and Alese Pechter, Dee Scarr, Marjorie Bank, Emory Kristoff, David Doubilet, Bill Lovin, Stephen Frink, Frank Fennell, Pete Nawrocky, Patricia St. John, Carl Roessler, Tom Campbell, Amos Nachoum, Patricia Scharr, and John Fine. These divers have brought an artistic as well as academic awareness of

*To learn about the marine environment, you can attend lectures
with guest speakers such as Jacques Cousteau.*

Photo: Mort and Alese Pechter –

Photo: Tom Campbell –

*Underwater photographers, such as Tom Campbell, raise worldwide
awareness of the problems in the marine environment.*

endangered living marine resources to millions of people of all ages. In 1993, Tom Campbell, a west coast-based underwater photographer, donated his dramatic images of environmental hazards along California's coast and had them made into a calendar. The non-profit project was sponsored by a local bank in Santa Barbara, and the proceeds were donated to marine sanctuaries and environmental awareness programs for children in California. Mort and Alese Pechter, Dee Scarr, and Pete Nawrocky have created award-winning multi-media shows on the marine environment for children, many of which they present as a free public service. These are just a few examples of how dedicated divers can make a difference.

NOTES

CONSERVATION FOR SCUBA DIVERS AND SNORKELERS

"Divers are indispensable to bio-DIVERS-ity."   **– John Regan, Scuba Instructor**

Divers have a personal "hands on" investment in the marine environment. Because their enjoyment, safety, and financial investment in diving depends on the aesthetic and biological integrity and the accessibility of dive sites, divers tend to be especially outspoken and protective of those sites. In fact, divers are often the first people to notice and report environmental problems, and that information is invaluable to marine scientists and management agencies.

Divers need to understand that there is an interlocking structure to the problem that encompasses matters that extend far beyond scuba and to communicate this information to their peers. Examining some of the other aspects of the problem will hopefully clarify some of these dimensions. Therefore, once divers understand the issues in marine conservation in general, they should seek to learn about problems in their local areas and in the sites where they dive.

If you are a diver, in addition to the avenues for academic education recommended in Chapter 2, you can obtain additional conservation insight and materials by networking with local environmental experts (from colleges, marine biology labs, aquariums, or museums), by inviting guest speakers to your dive club or dive shop, and by joining and supporting diving-related environmental organizations. (See Appendix 5) You and your dive buddies can make plans to participate in regional environmental projects and can integrate marine conservation activities, such as underwater cleanups (Chapter 17), into your open water dives.

MINIMAL IMPACT DIVING SKILLS

Divers with an appreciation and commitment to the marine environment practice "minimal impact diving" skills and "positive impact diving" behavior. Conservation must begin with improvement. Dive Leaders should demonstrate and help students fine tune their skills in buoyancy control, weighting, streamlining of the diver's body and equipment, and correct finning techniques, and teach environmentally responsible in-water etiquette.

BUOYANCY CONTROL

The most important skill which promotes "minimal impact diving" is buoyancy control. Divers with poor buoyancy control can inadvertently collide with and damage coral and other marine habitats. Even a seemingly insignificant brush against coral can remove its protective mucus coating, making it vulnerable to algae infestation, and subsequent fatal disease.

Buoyancy control is a skill that many divers, particularly novices, have trouble mastering. Therefore, Dive Leaders should demonstrate and constantly emphasize buoyancy control at all levels of training, beginning with the first pool session of the entry level course. A good introductory buoyancy skill is the "fin pivot,"

Underweighted divers often grab on to anything in sight for stability.

Photo: Dan Orr–

Photo: Pete Nawrocky –

Diver practicing buoyancy control by hovering horizontally in mid-water

Diver hovering vertically in mid-water

Photo: Pete Nawrocky –

Photo: Ric West

*Divers should use relaxed, controlled breathing in coral caves,
tunnels, and overhangs to avoid stirring up silt.*

in which the diver balances on her fins and rises and falls gradually by controlled inhalations and exhalations. As divers progress to more advanced diving activities, buoyancy control should be perfected.

Even though buoyancy control is taught at the entry level, constant practice is needed, preferably under a variety of conditions. An excellent training aid for divers is the PADI video, "Peak Performance Buoyancy." In confined and open water, buoyancy control can be practiced with an underwater obstacle course, such as the BTSI Diamond Reef Buoyancy Training System.

Buoyancy should be fined tuned by exercising proper breathing techniques. A slow, relaxed breathing pattern is imperative when diving in coral tunnels, caves, and overhangs to avoid stirring up sediment and disturbing delicate inhabitants.

NOTE: Make sure that you distinguish clearly between slow, relaxed breathing and "skip breathing," or breath holding, which can be extremely hazardous for divers and should be avoided.

PROPER WEIGHTING

A key to buoyancy control is correct weighting. Most novices and many infrequent divers wear too much weight and rely too heavily on the buoyancy compensator for depth control. A diver who is properly weighted, relaxed, and breathing comfortably should not have to rely excessively on the BC and fins for movement.

A diver who is underweighted will grab something for stability. An overweighted diver will flail and kick up the bottom, or resort to continuous inflation and deflation of the BC, known as "elevator diving."

To avoid these problems, take the time to understand the principles and variables of buoyancy control and make sure you are correctly weighted for each diving situation. Regardless of which technique you use to determine the correct amount of weight, you should be able to achieve neutral buoyancy, or hover motionless both horizontally and vertically within the water column. It is also important to recalculate and adjust your weight to compensate for changes in buoyancy, such as when going from fresh to salt water, when changing from an aluminum tank to a steel tank, or when changing to a thermal suit of a different thickness.

Changes in weight to adjust to new buoyancy factors should ideally be tested in a pool before diving in open water. In open water, divers should be properly weighted so that they can comfortably achieve neutral buoyancy and remain far enough away from marine life to avoid collision. There are many techniques and devices to learn and perfect buoyancy control, beginning with a simple fin pivot and progressing to more difficult buoyancy skills.

STREAMLINING

All divers should present an aerodynamically efficient, streamlined profile, which enables graceful gliding through the water. To achieve streamlining, the diver must be weighted correctly with arms close to the body and should use straight and fluid leg movements. Heavy ankle weights should be avoided, as they make proper horizontal trim difficult. If a diver absolutely needs ankle weights, such as for use with a dry suit, the minimum amount of ankle weight needed should be used.

It is also important to streamline equipment. Dangling gauges, hoses, and loose gear make moving through the water more difficult and present safety hazards to the diver as well as to the marine environment. Alternate air source stages, whether from an octopus or a pony bottle, should be attached to a holder on the chest in a readily visible and accessible position. Hoses and consoles should be secured to a customized Velcro fitting, a quick release buckle, or a clip attached to the BC. Underwater photographers and cinematographers must be proficient in methods of handling and harnessing bulky photographic equipment. You should not carry cumbersome equipment until you have mastered buoyancy control. The more skill- and equipment-intensive your diving activity, the finer your buoyancy control should be.

CORRECT FINNING TECHNIQUES

To promote minimal impact diving, you must practice and perfect your finning techniques. Incorrect finning is particularly harmful in coral reef environments. Poor finning can break coral and stir up clouds of sediment that can smother coral colonies. Before divers visit coral reef areas, they should be proficient in non-destructive finning techniques, such as the modified flutter kick (a flutter kick with smaller leg movements) and the "cave diver's kick" (legs bent up at the knee).

Streamlined equipment makes moving through the water easier for the diver and safer for the environment.

–Photo: Courtesy of the DACOR Corporation

Photo: Mort & Alese Pechter–

Dangling hoses and extraneous equipment can collide with fragile habitats.

With a correct finning technique, moving through the water should be fluid and effortless. Hands should generally remain close to the body. Regardless of the specific finning technique used, when diving close to coral, it is always advisable for a diver's legs to be somewhat higher than the rest of the body. Until buoyancy control is perfected, divers should adopt the "one meter" rule, which suggests keeping at least one meter (3 feet) clear of the coral. In a surge situation, divers should stay off the bottom and move with the surge, instead of flailing arms and legs or grabbing onto or crashing into coral.

For an enjoyable and non-invasive way of observing a reef up close, some diving leaders suggest the technique of "sculling." When sculling, the diver's body remains relatively motionless, while the hands alone (not sweeping arms) make small circular motions. Sculling helps a diver maintain a fairly stationary position by compensating for moderate water movement. A "lazy" one-legged kick may be sufficient to maintain trim with this technique. Unfortunately, many divers, particularly novices, have not mastered minimal impact diving skills.

Photo: Pete Nawrocky—

Sculling is a good technique to use near delicate habitats to avoid collisions.

POSITIVE IMPACT DIVING BEHAVIOR

A conservation-minded diver should demonstrate and impress upon fellow divers the importance of environmentally responsible behavior in and around the water. To promote underwater conservation:

1. Make sure that your diving skills are up to date and appropriate for the environment in which you are diving. If it has been some time since you have been underwater, do not hesitate to take extra lessons or a refresher course.

2. Always seek an environmental orientation before diving. The dive briefing should set the scene for environmentally responsible behavior, with emphasis on the area's specific pleasures and precautions. A good briefing should include information not only about weather and water conditions, such as tides, currents, waves, surge, visibility, temperature and thermoclines, but also a discussion of underwater topography, local marine life, and game laws.

3. Do not molest or alter underwater habitats, particularly in coral reef environments. Divers should not sit on, stand on, or break off coral or take coral for souvenirs. Underwater photographers and models should be especially mindful of their movements and behavior. Moving or removing rocks to set up a photograph can disrupt countless organisms.

4. Do not capture and take fish for salt water aquarium pets. This practice injures and kills fish and disrupts fish populations. Divers should not remove shells with animals in them or any other living marine organism. Skin and scuba divers should adopt the ethic: "Take only pictures, leave only bubbles."

5. Do not wear gloves in tropical dive sites. A bare-handed diver is more aware of and less stressful to coral. At dive sites where interaction with fish and marine animals is permitted, divers with bare hands will not as readily remove the creature's protective mucus coating by inadvertent rough handling. Coral and other marine creatures need their mucus coating for protection from disease and infection. If gloves are absolutely necessary to avoid injury from

Incorrect finning causes siltation, collisions, and creates drag.

Photo: Dan Orr –

Photo: Pete Nawrocky–

The flutter kick which divers are usually taught may be too forceful in fragile habitats.

rope burns or stinging organisms attached to your ascent line, wear gloves only when exposed to the hazard and leave them in your BC pocket at all other times.

6. Know, talk about, and respect local game laws. Even laws with which you may personally disagree should be followed while you work to have them changed. Divers should only take fish where underwater hunting is allowed, know the species, limits, and minimum allowable size that can be taken, and take only what you will eat. Violation of game laws can result in stiff fines and confiscation of the diver's equipment, but peer pressure can be an even more powerful deterrent.

7. Do not feed fish or other marine wildlife.

8. Do not molest marine animals, particularly turtles. Although they often appear docile and even playful, turtles can be traumatized by divers who ride them. Turtles are not fish. They are air-breathing reptiles. As such, they need to surface regularly. A diver who rides a turtle may be

asphyxiating it. Pregnant female turtles can abort their eggs after being traumatized by "playful" visitors.

9. Practice non-destructive boat anchoring. Many marine sanctuaries and protected areas around the world have installed permanent mooring buoy systems, such as the one designed by marine biologist John Halas. If a mooring buoy system is not available, dive boats should anchor on a sand or rubble bottom whenever possible. Knowledgeable captains and crews and Dive Leaders usually support this policy, understanding that preserving the environment is integral to their livelihoods. Divers should follow their example. Popular areas and sites can benefit from divers encouraging cooperation among diving professionals to establish and maintain mooring systems.

10. Never allow garbage, especially plastic, to be thrown improperly into the water. At the conclusion of a night dive, do not throw cyalume sticks overboard. Use the garbage receptacles on board your dive boat or at the dive site. If none are provided, make sure not to leave any trash, no matter how small, when you leave. To avoid the problem

A diver with poor finning and lack of awareness of his fins can stir up clouds of silt.
Photo: Pete Nawrocky–

of plastic debris entirely, do not bring any plastic disposable items on a dive. Divers should also carry a small collecting bag on every dive and use it to remove identifiable
debris from reef areas and other dive sites. Divers should
routinely recover cans, bottles, plastic items, and so on
from dive sites. Vessels should have a recycling bin for
recovered items. If diving from a beach location, bring a
bucket to use for litter during the day, and follow the
beach's recycling rules. Make sure that cigarette butts are
not left in or around the dive site. The filters are not
biodegradable and may be eaten by birds that mistake
them for food. It should be the active goal of every diver to
leave the environment cleaner each time he or she leaves
the dive site.

*A good rule of thumb is to
remain at least a meter
(three feet) away from
coral until you have
perfected your
buoyancy control.*

— Photo: Tom Mount

Divers should configure their equipment close to the body so that it doesn't collide with marine habitats.

−Photo: Courtesy of DACOR Corporation

11. Check that all dive gear is properly secured before you enter the water, so that nothing will drop or break off.

12. Report violations of debris laws. Under the terms of the MARPOL Annex V Law, illegally discharged debris can result in serious legal penalties. Divers should identify and report underwater areas where large amounts of debris have accumulated to the Center for Marine Conservation in Washington, DC. This information is incorporated into a computerized database and used by local and federal government agencies and private industry to confront and penalize offenders.

13. Take special care when diving in underwater caves and caverns. Avoid crowding into the cave, and unless you are on a scientific expedition, do not spend too long in the

cave. Air bubbles collect in pockets on the roof of the cave, which can harm the delicate creatures living there. Always move carefully and avoid forceful exhalations, which can cause siltation.

14. When planning a dive trip, incorporate a reef monitoring or other conservation program into your trip. (See Chapter 18.)

15. Learn underwater photography and or videography and use these skills to document environmental damage. Underwater slides and videos can be used by resource managers to monitor and enforce conservation laws, and in educational and outreach programs.

16. Do not buy endangered species. Do not buy tropical marine fish or other live reef animals, including corals, to keep in an aquarium unless you are certain that the species are being collected and marketed in a sustainable and responsible way. International trade in wildlife is now controlled in over 100 countries by the Convention on International Trade in Endangered Species of Wild Fauna and Flora (CITES).

Careless behavior by underwater photographers can destroy delicate ecosystems.

–Photo: Pete Nawrocky

Photo: Reef Relief –

Untrained snorkelers can harm coral.

17. Snorkelers should observe the same rules and precautions as scuba divers. Although proof of certification is not required for snorkeling in most areas, environmentally responsible snorkelers should take a skin diving course and learn techniques to minimize impacts to the environments they visit. Untrained snorkelers can cause considerable damage to fragile habitats and their inhabitants.

A helpful reminder, the mnemonic ECO-CHAMPION, summarizes the key points of minimal impact diving and positive impact behavior. (See Figure 18)

BE AN ECO-CHAMPION

Ensure that you stay far enough away from coral and other marine habitats to avoid collision.

Control your buoyancy.

Obey game regulations and marine debris laws.

Calculate and wear the correct amount of weight.

Have your equipment properly configured close to your body.

Attach your boat to a mooring buoy or anchor on a sand bottom.

Maintain a streamlined body position.

Practice proper finning techniques.

Interact with fish and marine animals by gentle, non-threatening behavior, and do not offer food.

Only take pictures, not souvenirs.

Never wear gloves around coral.

Dr. Sylvia Earle inspecting
a contaminated beach
in Kuwait during
the Gulf War

−Photo: courtesy Dr. Sylvia Earle

IDENTIFYING AND RECORDING DATA

A conservation-minded diver is always on the lookout for environmental alterations and damage. In order to be able to recognize and report problems in a particular marine environment, one must first become familiar with that environment's normal components and activities. Even scientists must establish a database from which to work. Marine biologist Dr. Sylvia Earle, directed by NOAA to assess the damage in Kuwait from the burning of oil fields during the Persian Gulf War in 1991, recalls the difficulty of her mission: "Despite all of our modern technology, it was virtually impossible to assess how many marine species had been lost, because no one had ever established a database against which we could measure the damage!"

This does not mean that to support conservation you have to conduct a comprehensive scientific study of every dive site you visit. Nor do you have to be fluent in the Latin names of each genus and species. But learning to make general identifications of typical marine plants and animals is an excellent idea. Since fish are among the most interesting creatures that divers are likely to observe, even broad terms that identify fish by their habitat can be useful:

Anadromous: Anadromous species spend part of their lives in the ocean but migrate to fresh water to spawn. Striped bass and salmon are examples of anadromous fish.

Pelagic: Pelagic means "open ocean," and pelagic fish live at or near the water's surface in the open ocean. They are usually large fish that migrate long distances. Examples include swordfish, tuna, and many species of sharks.

Coastal pelagic: Coastal pelagics live closer to the coast than true pelagics, but still swim at or near the water's surface. Examples are mackerel, anchovies, and sardines.

Billfish: These fish have elongated, spear-like protrusions at the snout. Swordfish, sailfish, and marlin are examples. Billfish are usually pelagic.

Groundfish: Groundfish is a collective term for fish which live on or near the sea floor. Groundfish include a variety of bottom fishes, such as cod, haddock, and pollock, and flatfish, such as flounder.

Reef fish: Reef fish live in or around reefs for most of their lives. "Reef fish" is usually used collectively for snappers, groupers, porgies, grunts, and other associated species. *Reef Fish Identification* by Paul Humann (see below) is a highly recommended guide.

Invertebrates: Invertebrates are shellfish and other animals without backbones. This group includes lobsters, clams, shrimps, oysters, crabs, sea urchins, scallops, and many other species.

There are numerous slides, photos, posters, films, and reference books by which divers can familiarize themselves with local marine life. The series of books by Paul Humann, *Reef Fish Identification, Creature Identification,* and *Coral Identification,* are excellent references books for marine life in Florida, the Bahamas, and the Caribbean. The week-long REEF Fish Identification Program, also developed by Humann, is highly recommended. (See Chapter 18.) More affordable alternatives are *"Caribbean Reef Encounters,"* a tutorial video accredited course by Delphin Productions, which provides an introduction to marine life for non-scientists. Another excellent learning tool is David K. Bulloch's book, *The Underwater*

–Photo: Paul Humann/REEF

Divers can record and report underwater changes and damage.

Naturalist. (See recommended readings in Appendix 4.) This is essentially a layman's guide to marine life classification and identification, with instructions for measuring water temperature, dissolved oxygen, and salinity, and for collecting specimens.

When diving, try to learn as much as you can about the site beforehand from academic materials and local diving professionals. Ask your Divemaster to include a brief discussion of marine life in his/her dive orientation. Most important, bring a local marine life identification book and an underwater slate. The slate can be used to record broad categories of marine life and activities, as well as to record general quantities of species observed. If you are at a loss for names, try to sketch or describe what you see. At the conclusion of a dive, divers can compare and discuss their observations with the Divemaster. Marine life identification is an excellent activity which can be done independently or as part of an open water scuba course.

IDENTIFYING AND REPORTING ENVIRONMENTAL DAMAGE AND ABUSE

Once you have a general knowledge of the habitats and marine life at your dive site, watch for unusual events or changes. Divers have been extremely helpful in sighting and reporting such problems as sea urchin devastation, coral diseases, marine mammal strandings, and violation of marine debris dumping laws.

Use your underwater slate and compass to record your observations and location. Even if you are unable to identify a species, write down (or draw) its size, color, notable characteristics, and the extent of the problem. Find out beforehand the appropriate management agencies for reporting your observations. If you are diving specifically to locate suspected environmental damage, use buoy markers to help officials make follow-up verifications of your findings. Never attempt to bring up hazardous materials unless you are adequately trained and equipped.

Evidence of beach pollution, such as garbage slicks, brown or red tides, and fish kills, should be reported to the Coast Guard, EPA, DEP (Department of Environmental Protection), or Marine Police. Chapters 19 and 21 provide information on how to be an effective activist for environmental issues and include tips for writing letters to regional councils and Congress, for testifying at public hearings, for organizing a grassroots network, and for alerting the media. This material should be incorporated into open water diving courses and dive briefings and de-briefings.

Underwater photographers may want to incorporate photo documentation of environmental damage into their dives. The Center for Marine Conservation is particularly interested in obtaining photos of plastics and trash wrapped around coral and wildlife and pictures of animals using debris for shelter or homes (an octopus living in a can, coral growing on a bottle, etc.). This type of photo documentation can also be an excellent project in an underwater photography specialty course.

CONSERVATION FOR RECREATIONAL BOATERS AND FISHERS

"A survey of the ocean out of sight of land is a good way to get some perspective on your place in the grand scheme. That limitless expanse edging the heavens all around, relieved only by swell, ripple, and wave, is a definition of eternity." — *Capt. Ira Barocas*

Like divers, recreational boaters and fishers need a healthy environment for the safety and enjoyment of their activities. On a large scale this requires that boaters become involved in legislation and enforcement to protect the environment. On the personal scale, it is also necessary for each boat owner, operator, and crew member to take responsibility for his or her behavior on, around, and in the water. Recreational user groups usually do not create major marine environmental stress, but incorporating conservation into boating and fishing can greatly reduce accidental damage.

CONSERVATION FOR RECREATIONAL BOATERS

Boating is a popular sport, and there are approximately eighteen to nineteen million recreational vessels in the United States alone. "Recreational boats" encompass a large variety of craft from large yachts to jet ski craft. Yet there are currently no reliable quantitative data of the impacts this group has on aquatic habitats and wildlife.

Ira Barocas, a professional diver and boat captain, has been navigating waters throughout the Atlantic and Pacific Oceans and Caribbean Sea for almost two decades. In the following paragraphs, excerpted from articles in *Sources* magazine, Capt. Barocas offers his perspective on conservation for boaters:

The simple concept of buoyancy is all that keeps us aloft over the deep; our good sense and elemental technology enable us to use this living element, the sea, for our liveli-

Boaters must observe responsible conservation laws.

hood and pleasure. The immensity of the sea, we've learned unfortunately, is no guarantee of invulnerability to destruction. Our incredible ingenuity and a tradition of abuse and selfishness are fast overtaking the viability of our most precious resources. While the problem is global, the solution is individual: it is for each of us, especially we, the "water people," to insure that our impact on our chosen element is not just benign, but beneficial.

Admittedly, recreational boating is just a sliver of the forces that act on the environment, and there are laws that purport to protect the sea, littoral zones, aquifers, and the life they support. The International Convention for the Prevention of Pollution by Ships (MARPOL) prohibits dumping of any litter within twenty-five miles of a coastline, or any plastic anywhere at sea, and is joined by volumes of federal, state, and local laws and regulations in our country and throughout the world.

Coastal Resources Management laws, catch quotas, recycling efforts, and the Ocean Dumping Act of 1973 that places restrictions on dump sites, creates sanctuaries, and requires EPA permits for offshore dumping, are in place and are

Photo: John Halas–

Boats should be attached to a mooring buoy.

The mooring buoy is attached to the bottom so anchors do not have to impact coral.

Photo: John Halas–

enforced as much as possible. There are also the Fish and Wildlife regulations and other conservation efforts and agencies. But the sad fact of the matter is that it is not a problem that laws and vigorous enforcement can satisfactorily address.

Only five percent of the tens of thousands of ships docking annually in our nation's ports leave any trash ashore. Storage and removal is costly if and when available. What happens to the approximately 3 pounds of trash per person per day (a U.S. Navy estimate) that a ship at sea generates? What about the 50 million pounds of plastic packaging lost yearly by commercial fisheries, or the 300 million pounds of nets, floats, and buoys according to a National Academy of Science estimate?

Policies that we establish aboard our boats, and in our teaching environments affect everyone. There are things boat owners, captains, and crews can do, besides talk about it, to help foster the kind of attitude necessary to ameliorate the problem we face:

Obey the MARPOL Annex V Law, and post a MARPOL sticker on your vessel. Never allow trash to be discarded overboard. Nothing at all, not a tissue, cigarette butt, match stick, bottle cap, sandwich remains, anything which was brought aboard, should go off the boat while it is out. Bring it all back and dispose of it correctly ashore. Provide appropriate containers for recycling glass, plastics, and paper. Keep some shears near the containers so that bulky boxes, six-pack rings, and other items can be "treated" before discarding ashore. Consolidate less than full sacks of trash before discarding ashore. Enforce this among your passengers and especially crew.

Buy foodstuffs in bulk to reduce packaging refuse, and use containers for serving and storage aboard. A bag of chips is the same as a bowl of chips, and the bowl can later be washed out and reused and the chips refilled from a larger container. Minimize the use of disposables, especially among crew. For example, have labeled drinking containers. A mug with a name on it is an honor for "regulars."

Don't buy or use or allow aboard anything Styrofoam. Offsounding on extended voyages, only food leftovers should be discarded. Other refuse should be saved aboard until landfall. Work out your stowage plan in advance. This is possible for food, waste oil, and all trash.

If you are transporting divers, check carefully for loose gear, poor placement, faulty attachments, etc. Breakage and loss contribute to reef and site destruction. Fish can die as easily from ingesting a plastic snorkel keeper as from eating a plastic bag. Susan Walker, Director of the Saba Marine Park, NA, tells the story of a particularly friendly large tiger grouper, "Sweetlips," who apparently died after ingesting a lost dive knife. The animal was seen with the blade protruding from her belly after a visiting diver reported losing the knife on the grouper's habitual site.

If you have a toilet onboard, it has to comply with the current standards, and essentially those standards are most easily satisfied by the term: no discharge. In the days of wooden ships and iron men, the traditional cedar bucket was supplemented by a trip forward to the carving at the prow of the vessel. There the sailor would "anoint the waves," climb back aboard, and go on about the daily routine. From these trips f'ord came the term "head" in its usage as nautical for toilet. This quaint practice, ripe with more than just history, however, is no longer allowed. There is no discharge allowed within three miles of shore anywhere in the United States, period. In fact, many states, with the permission of the Coast Guard and the EPA, have established even more extensive no discharge zones covering whole stretches of coastline, some large bays and harbors, lakes and tributaries, and other waterways. Local authorities are the best guide to where you can legally discharge waste in your boating area. But the truth is that the only really acceptable system that you might install aboard would be a holding tank, or as it officially known, a Type III MSD.

In EPA designated no discharge zones, for instance San Diego Bay, the entire state of Michigan, or the inland waters of Maine, pumpout facilities are relatively convenient and available. In other states, vessels may have to get outside the limit and then pump overboard. This requires a system that's plumbed with a Y-valve that allows waste matter to be directed to either the holding tank, or an overboard discharge. Offshore vessels generally have a flow-through system for operating outside the limits. Some of the larger ones are equipped with onboard incinerators. In any case, a Y-valve must be disabled in port or within a no discharge zone by either locking, removal, or some other semi-permanent means to prevent accidental overboard discharge.

-Photo: John Halas

Coral which has been damaged by a boat anchor.

There are two possible alternatives to the holding tank, except in the declared no discharge zones–the Type I and Type II devices. Both of these rely on some form of waste treatment that will render the effluent less environmentally harmful. Vessels larger than sixty-five feet cannot use Type I devices, and inspected passenger vessels must have holding tanks. The two systems chemically treat and macerate the waste so that the discharged product has a coliform bacteria content of less than 1,000 parts per 100 millimeters for Type I systems, or less than 200 parts per 100 milliliters for a Type II system. In neither case are there visible solid wastes.

Voluntary compliance with the strictest standards now seems the most prudent course of action to protect the environment and to retain the important leadership role which boat captains play in conservation.

In addition to Captain Barocas' recommendations, here are some more helpful hints for boat owners and operators:

1. Never anchor on a fragile habitat. If a mooring buoy is not available, use a sand or other non-living bottom. Keep your

anchor chain as short as possible to prevent it from dragging on reef areas. When retrieving your anchor, motor towards the anchor to prevent dragging.

2. Exercise special care when navigating in coral reef areas.
3. Avoid using toxic TBT paints on your boat. A safe alternative is anti-fouling paint which can be purchased at local marinas and boating stores.
4. Use canvas boat covers instead of disposable plastic covers. Canvas is sturdier and less harmful to the environment.
5. Make sure your motor does not leak gas or oil into the water. Do not drain any engine fluids into the water.
6. Place a bilge "pillow" (an oil-absorbing sponge available in marina stores) in your bilge to remove oil from your bilge water.
7. When cleaning your boat, use a non-phosphate detergent and a scrub brush.

If you are a customer on a boat, you have the right to express your conservation concerns. If you are booking a trip on a commercial boat, check in advance as to the company's environmental policies, particularly about anchoring and discharge of sewage and trash.

CONSERVATION FOR RECREATIONAL FISHERS

"Marine fish are the canaries in the coal mine of the oceans. It is up to us to ensure their complexity and abundance for future generations."

– David Allison, Center for Marine Conservation

Recreational fishing is another activity whose roots are entrenched in centuries of natural history and cultural mores. For many, to reap the bounty of the sea is not only an adventure and technical challenge, but an affirmation of one's physical prowess and survival skills. In the light of the global crises of overfishing and unsound practices of some commercial entities, even recreational fishing has recently met with opposition. If fishing laws are obeyed and ecological etiquette is practiced, however, sport fishing can be compatible with conservation.

Fishing organizations usually write guidelines for personal conduct. In addition, NOAA has created an "ethical angling" program whose motto is: Commit yourself to ethical angling, the future of your sport depends on it. Pass it on!!!

Help fish stocks increase, through catch and release.

Limit your take, don't always take your limit.

Observe regulations and report violations.

Only keep fish for trophy or dish.

Escape tradition, try a new catch in the kitchen.

Get hooked on fishing's thrill, not alcohol or drugs that kill.

Bring all garbage in, don't teach it to swim.

Captain your boat, practicing safety afloat.

Show courtesy and respect, others rights don't neglect.

Share what you know to help your sport grow.

Additionally, when fishing from a boat, follow the conservation guidelines for boaters, especially those concerning plastic and trash disposal and anchoring:

1. Obey the MARPOL Annex V Law. All fishing boats should post a MARPOL sticker which stipulates the overboard dumping regulations. Never throw fishing line or other plastics overboard. Keep a garbage receptacle on board, keep it covered, and make sure that everyone on board uses it. If you dispose of your garbage at a marina, follow their recycling rules.

2. Never litter, and do not allow others to litter. Avoid using disposable plastic products or dumping cigarette butts in the water while fishing.

3. Do not discard fishing line. Cut used line into small pieces and dispose of it properly. Contact the nearest fishing line recycling center in your area.

4. Always obey local fishing laws. Do not violate rules regulating allowable catch size and quantity, fishing gear, seasonal or area limitations, and endangered species.

5. Do not destroy "trash" fish. Each species plays an important role in the ecosystem. Always carefully release all unwanted species.

6. Do not disturb habitats. Stay on designated paths and stay well off sand dunes, sod banks, coral, and other fragile areas.

7. Catch and release. Preserve future stocks by releasing small fish and fish you cannot immediately use. Use barbless single hooks for fish you plan on releasing.

8. Participate in tag fishing. Tagging and releasing fish help scientists learn more about growth rates and migration routes of valuable species, such as tuna, swordfish, marlin, sharks, and striped bass. To participate in fish tagging expeditions, contact the American Littoral Society (listed in Appendix 3) or the National Marine Fisheries Service (listed in Appendix 2).

9. Make sure that passive fishing gear (such as traps and gill-nets) are being attended and retrieved regularly.

<u>NOTES</u>

CONSERVATION ACTIVITIES
BEYOND THE WATER

"When it comes to the future, there are three kinds of people: those that let
it happen, those that make it happen, and those who wonder what happened."

– Carol Christensen

It is critical that divers, fishers, boaters, swimmers, and other individuals who enjoy aquatic recreation become leaders in marine conservation, not only in the water, but beyond. Immediate access to the aquatic world gives these individuals an excellent vantage point for identifying and monitoring environmental problems, as well as the ability to offer practical input for their solutions. Even though recreational user groups do not contribute generally to marine environmental problems, there are things they can and should do to help. Conservation needs role models. By setting a personal example, people can inspire their friends, peers, family, legislators, and influence their habits.

STAY CURRENT ON ENVIRONMENTAL ISSUES

The material presented in this book is only a starting point for marine ecology and conservation awareness. There are many other ways to gain comprehensive knowledge and stay current on environmental topics. As mentioned in the introduction of this book, college courses and lectures are prime educational sources.

CLEANUPS

One of the most effective ways you can help with the problem of marine debris is to organize and participate in shoreline and underwater cleanups. Beach and fresh water cleanups, held in many geographic locations, are extremely popular and effective events which can involve non-divers as well as divers. Every year, volunteers clean up thousands of tons of debris from coastlines and lakes around the world. Cleanups not only reduce the amount of trash, but they help identify and penalize offenders of the MARPOL Annex V Treaty. (See Chapter 17 for a comprehensive plan for organizing and conducting cleanups, with tips on sponsorship, publicity, and safety protocol.)

Cleanups can center around known problems in your area. For instance, because discarded fishing line is a problem in Florida, cleanup volunteers calculate the total amount of line found in the state. Some states conduct recycling demonstrations in conjunction with cleanups, transforming plastic debris into useful items like park benches. CMC has developed Data Identification Cards to help record and categorize debris collected during cleanups. When these cards are returned to CMC, the information is entered into a large database and is used by government agencies, legislators, industry officials, and local communities. The Data Cards help trace debris to common sources, such as specific cruise lines and manufacturers. Offenders can then be confronted with the information and targeted for reform. For example, data showing the amount of Styrofoam on Texas beaches a few years ago convinced Conoco Corporation to prohibit Styrofoam products on their oil rigs.

In addition to the annual CMC cleanups, it is important for individual groups to conduct underwater cleanups of local dive sites. According to statistics from PADI Project AWARE, 17,235 divers participated in beach and/or underwater cleanups worldwide in 1993. This only consists of information received at PADI from its members, and does not include divers who participated in cleanups which were not reported to PADI.

Divers, boaters, fishers, and swimmers should make a cleanup part of all open water activities. If you are with a group, consider offering a prize to the diver who collects the greatest amount or the most unusual items of trash. Encourage your peers to make a habit of always carrying a trash bag in their dive bag, in case a proper trash receptacle is not available at a dive site, and to leave underwater and topside dive sites cleaner than they found them.

RECOVERY OF GHOST FISHING GEAR

Ghost fishing gear, such as discarded monofilament fishing line left in the water, can be deadly to marine life. According to coastal cleanup reports, monofilament line is responsible for over one third of all wildlife entanglement incidents. With proper training, divers can recover ghost fishing gear and dispose of it properly. Search and recovery of ghost fishing gear can be a valuable exercise in advanced diving classes. Berkley, a fishing tackle manufacturer, located in Spirit Lake, Iowa, offers a nationwide recycling program for monofilament line.

HELP REDUCE MARINE POLLUTION AND RECYCLE

Visitors to aquatic environments should follow and encourage others to follow proper waste disposal protocols, not only at dive sites, but in everyday life. (See Chapter 22) Many household products, such as paint and nail polish removers, detergents, motor oil, etc., contain chemicals which can pollute ground water. Check with your sanitation department or a local conservation agency for instructions for the disposal of caustic and toxic materials. Check to

Plastic can be recycled into many useful items, even park benches.

–Photo: Plastics for Progress Alliance

see if there are, or help establish, facilities in your community for recycling glass, metal, paper, and plastic.

Buying recycled products may cost a little more, but it is an important avenue for conservation. Of course, it is also helpful to decrease your overall use of plastic and packing material. If plastic six-pack holders are cut up before being discarded, they cannot strangle marine wildlife. In response to public and government demand, manufacturers are currently developing alternate forms of degradable plastic. Lastly, you can elect not to patronize companies whose actions and policies are environmentally irresponsible.

JOIN AND SUPPORT ENVIRONMENTAL ORGANIZATIONS AND CLUBS

Everyone should join NGOs (non-government organizations) in local and regional areas. Environmental organizations are important for keeping up to date on marine conservation issues, and some will provide educational materials gratis or at a minimal fee. (See Appendix 3.) Some national organizations may have a local or regional chapter in your area.

Contacting and getting involved with environmental clubs and organizations is also an effective vehicle for individuals to voice their needs and concerns to industry and the government. These groups are often influential in creating and enforcing local laws regarding sewage treatment, pollution, water quality, chemical dumping, fishing, zoning, and coastal management. In 1981, for example, hundreds of letters from environmental group members in California, which were sent to congressional representatives and to the U.S. Department of Commerce, helped get Point Reyes and the Gulf of the Farallones designated as national marine sanctuaries.

In 1994, fifteen environmental groups, from both the diving industry and other conservation communities, formed a coalition to help protect and strengthen environmental laws, particularly the laws which are due for reauthorization by Congress. The coalition members are: American Oceans Campaign, Center for Marine Conservation, Defenders of Wildlife, Environmental Action Foundation, Friends of the Earth, Greenpeace, League of Conservation Voters, National Audubon Society, National Parks and Conservation Association, National Wildlife Federation, Natural Resources Defense Council, Sierra Club, Sierra Club Legal Defense Fund, The Wilderness Society, and Zero Population Growth.

Besides legislative action, some environmental clubs and organizations organize and conduct ecological travel and vacation seminars and programs. Many areas in the U.S. and abroad now offer opportunities to work at everything from recording whale behavior and marine census-taking to reef restoration. These field programs are a premier opportunity to work with and learn from leading experts. (See Chapter 18.)

HELP ESTABLISH STRANDED MARINE MAMMAL AND TURTLE NETWORKS

The Center for Marine Conservation is an excellent resource which obtains and publicizes marine mammal and sea turtle regulations and establishes stranding networks in local areas. CMC has

It is important to protect habitats for endangered species.

–Photo: Mort and Alese Pechter

*There are many avenues
for marine education
for children*

–Photo: Mort & Alese Pechter

compiled a list of emergency phone numbers and instructions for people who find marine mammals or turtles in distress. This information should be posted in aquatic sports facilities, as well as publicized to the general public. In many areas, local citizens have formed volunteer teams which work with the Coast Guard to maintain stranding networks.

SUPPORT MARINE SANCTUARIES AND PROTECTED RESOURCES

NOAA has developed a program of "interpretive enforcement" which is currently used in the national marine sanctuaries. (See Chapter 11.) Interpretive enforcement is a positive, supportive way of educating all sanctuary visitors including boaters, divers, snorkelers, and fishers and seeks to prevent abuse to the sensitive marine environment.

Recreational boaters, dive boats, and fishing boats in the Florida Keys, for example, are greeted by a NOAA patrol boat as they enter the NOAA marine sanctuary. After the NOAA official extends a friendly welcome, the group is given a waterproof packet of information containing tips to insure their safety as well as the safety of the marine environment. Visitors to the sanctuary are reminded not to remove or stress corals and their inhabitants and are advised as to the allowable game that can be taken. The NOAA patrolmen will gladly answer questions and lend assistance if needed.

Marine sanctuary visitors should stay current on NMSP activities and policies, support the NMSP, and educate others about the program's methods and goals. The NMSP is also an excellent source for environmental teaching materials.

SHARE YOUR KNOWLEDGE WITH OTHERS, PARTICULARLY CHILDREN

One of the most important ways we can actively promote marine conservation is to share our knowledge and enthusiasm with others, particularly children. The knowledge and attitudes we impart to the younger generation will profoundly affect our plan-

The Ocean Pals Poster contest teaches children to protect the marine environment.

Photo: Mort and Alese Pechter –

et's future. Marine conservation presentations for children can be made at public schools, libraries, aquariums, and community centers. "You Can Make A Difference" (available from NAUI) is an entertaining and informative marine conservation video for children, hosted by Dr. Sylvia Earle and news personality Storm Field.

There are many books which teach children environmental awareness and conservation, such as *What's in the Deep?* by award-winning photographers Mort and Alese Pechter (Acropolis Books, LTD, Washington, DC), *Touch the Sea* and *The Gentle Sea*, by Dee Scarr, and *Snorkeling for Kids of All Ages* (published by NAUI). The Center for Marine Conservation and PADI have public education programs and support materials for children. There are a number of academic and hands-on environmental programs for children, including Sea Camp (an environmental center in the Florida Keys), the Audubon Society (which hosts environmental camps in various areas), and Ocean Pals (Headquarters in Rye, NY). Appendix 5 lists organizations which provide environmental materials for educators.

ORGANIZING AND CONDUCTING SHORELINE AND UNDERWATER CLEANUPS

*"At no time in history has attention to the marine debris problem been greater.
We must use this momentum to confront the problem areas and eliminate their
contributions to the marine debris problem."*

– Kathy O'Hara, Director CMC International Coastal Cleanup

One of the most effective ways people can respond to the problem of marine debris is to organize and participate in shoreline and underwater cleanups. Cleanups not only reduce the amount of trash, but they help identify and penalize offenders of the MARPOL Annex V Treaty. Since 1988, cleanups have produced extensive information on the types and quantities of marine debris found in and around America's waters. Data recorded during cleanups can be sent to the Center for Marine Conservation in Washington, DC, where it is entered into a large database and used by government agencies, legislators, industry officials, and local communities. For example, the data collected from cleanups in 1986 was used to help pass the MARPOL Annex V Treaty and several state programs, such as the California Marine Debris Action Plan.

NOAA Marine Debris Office and the Center for Marine Conservation, both located in Washington, DC, organize and lend support to annual coastal cleanups. These cleanups involve thousands of volunteers who join together for one day to collect trash from beaches. Last year, 200,000 volunteers in thirty-five states, and U.S. Territories, and sixty countries collected 850 tons of debris in one day. Participants found more than 1,000 items with labels from forty-eight countries. Cleanups affirm how far trash can travel, as evidenced by cleanups in the Gulf of Mexico, where plastic items are found from as far away as Japan, Bulgaria, France, and Antarctica.

An annual international underwater and shoreline cleanup event debuted on September 16, 1995, and is scheduled to continue every year. The global cleanup is organized by the Underwater Cleanup and Conservation Monitoring Program (UCCMP), a coor-

A cleanup should have a dramatic theme.

OUR OCEAN. IT'S DROWNING.

Stow your trash and prevent marine debris.

—Photo: CMC

dinated effort of the diving community, federal agencies (EPA, NOAA, and USCG), and the Center for Marine Conservation. The goals of UCCMP are to remove debris from the underwater environment, educate the public and the scientific community about what is found, and influence policy decisions related to the handling of human-made debris that impacts marine and freshwater environments. Through large-scale cleanups and data assessment, UCCMP is seeking to learn to what extent submerged trash is killing wildlife such as marine mammals, turtles, and fish below the surface, and how different types of debris are altering marine ecosystems.

Divers as well as non-divers are welcome and encouraged to participate in the UCCMP international underwater and coastal cleanup, which include contest, prizes, and an underwater photo shoot-out (cleanup activities vary, depending on country, city, and cleanup site). For more information contact CMC at (804) 851-6734 or (800) CMC-BEACH.

How to Organize and Conduct a Cleanup

In addition to these large scale annual cleanup events, it is important for volunteers to help clean up local sites on a regular basis. To be most effective, shoreline as well as underwater cleanups should have good leadership and follow a well-organized plan, with an emphasis on fun and safety. The following is a step-by-step plan, developed by the author, for organizing and conducting a successful cleanup:

I. DECIDE WHO WILL BE THE MAIN SPONSORS OF YOUR CLEANUP

There is a lot of truth to the old adage, "strength in numbers." Try to involve several key people and groups in your cleanup. Good choices for cleanup sponsors are one or more dive clubs, dive shops, dive training agencies, boating groups, fishing groups, environmental organizations, a local school or university, or a local community center. Two years ago, with the help of two staff members from PADI Project AWARE and The Center for Underwater Research and Exploration (CURE), I created "Divers For A Cleaner Planet," the first inter-agency environmental alliance. For our first pilot project, NAUI, PADI Project AWARE, and CURE conducted a cleanup in New York City's Central Park Lake, an outstanding event which drew participants from all over the world and received international media coverage.

II. DELEGATE RESPONSIBILITIES

As competent and energetic as one person may be, it is best to share the work. Make sure that responsibilities are clearly defined and agreed upon by your "Executives," or key staff members and put in writing in the initial planning stages. These staff members may then appoint their own committees to help with their allocated responsibilities.

III. CREATE A THEME OR A MESSAGE AND A LOGO

Try to give your cleanup an attention-getting and relevant theme. For example, The Center for Marine Conservation launched its national cleanup campaign with the slogan, "Our Ocean–It's Drowning!" For a future cooperative cleanup, I would choose Atlantic City, New Jersey, and involve local casinos as sponsors. Gambling chips redeemable at the casinos could be hidden in the sand as part of a treasure hunt during the cleanup, and a striking theme could be, "We Can't Afford to Gamble on the Environment!" Another newsworthy cleanup site was the Statue of Liberty in New York City, where the theme was, "Freedom to Dive Clean Water Again."

Even if you cannot choose a high-profile cleanup site, you can still create a dynamic logo. A dramatic statement grabs headlines and attracts participants and news media.

IV. SOLICIT DONATIONS

Volunteers may be free, but cleanups are not. You will need money for advertising, equipment, mailings, refreshments, raffle tickets, souvenir tee shirts, etc. Ask local organizations and businesses for cash donations. If your solicitations of money are not successful, ask for merchandise or services. Donations in the form of free food, hotel accommodations, advertisement space, prizes, parking privileges, air fills, transportation, and garbage disposal service may actually be more useful than cash.

Cleanups can be conducted in freshwater and urban environments, as well as along coastlines. Pictured: Cleanup in New York City's Central Park Lake

–Photo: Pete Nawrocky

Photo: CMC –

Cleanups can involve entire families, even young children.

V. Seek Publicity

The attendance of your cleanup directly depends on how well you publicize it. There are a number of ways to get your message out, including mailings and posting fliers in local stores, community centers, dive shops, and aquatic clubs. Advertisements and press releases in local newspapers as well as in sports magazines and community newsletters are very effective forums. Hopefully your cleanup will be a non-profit event, (and profits beyond your operating expenses can be donated to the local cleanup facility or put into a fund for future cleanups) and as such will be eligible for a public service announcement. Depending on where you live, this may allow you free radio and cable television announcements and free newspaper ad space. However you choose to publicize your cleanup, make sure you include all the necessary details (date, time, location, parking arrangements, what equipment to bring, if there will be refreshments, prizes, other activities, and name and phone number of the contact person).

You also want publicity *during* your cleanup. Realistically, no group can completely eliminate all the debris from an entire

-Photo: Scott Jones/courtesy PADI

Cleanups should encourage teamwork

beach, lake, or river in one day. Your cleanup is important, therefore, not only to beautify the area as best you can, but to make a dynamic statement. You want as many people as you can to see that divers are concerned individuals who care about the underwater environment.

The best way to procure the most exposure for your environmental statement is by rallying the press. You do not have to be a public relations professional to get television, radio, newspaper, and magazine coverage. The press is always looking for exciting, offbeat, and timely happenings. Scuba divers are always fascinating to the public, and what could be more timely than environmental progress in action?

Prepare a simple press kit and send it to as many news/feature editors as possible at least two weeks before your event. In the press kit, state the details of your event, your goals, and how many people you expect to participate. If you have invited any celebrities or politicians to the cleanup, mention them by name. Include photos of past cleanups. Have volunteers call a few days before the cleanup to confirm whether a news crew has been scheduled to cover your event. Regardless of how many cameramen do attend, appoint your own qualified volunteer as the "Official Event Photographer." Photos and/or video are a valuable addition to your files.

Press people will be more likely to attend your cleanup if you have a catchy theme and/or choose an important or unusual site. For example, for the "Divers For A Cleaner Planet" cleanup, we specifically chose Central Park Lake with publicity in mind. Central Park Lake is an instantly recognizable landmark in New York City. It is also located in the middle of a highly commercial urban area, a most unlikely destination for underwater activities. But what really drew the swarm of press to our event was the fact that Central Park Lake was a 107-year-old body of water in which no scuba diver had ever been allowed before, so no one knew what might be retrieved from the bottom. The possibilities were intriguing!

VI. SURVEY THE AREA BEFOREHAND

Several months before your cleanup, the organizers of your event should visit the site and do a thorough survey. You may have to do this several times. Take photos and notes of everything. Obtain or make a map of the area. After consulting with the local authorities, decide and agree upon what the boundaries of your cleanup will be, the easiest and safest entry and exit points for the cleanup, what topside and underwater conditions will be encountered (temperature, depth, water quality, visibility, currents, etc.). Make sure that you know what kind of debris you may encounter, and if it is safe to handle. If you are planning an underwater cleanup, have your dive leaders make a supervised orientation dive prior to the day of the cleanup. When you conduct a survey, make a "mental movie" of exactly how your cleanup day will proceed from start to finish.

VII. HAVE THE NECESSARY PAPERWORK READY

Imagine having a few hundred volunteers loaded with gear arrive for a cleanup only to find out that there is no place to park their vehicles? To avoid problems like this, prepare all the necessary paperwork in advance. This includes parking and facility entry permits, directions and a map of the site, registration lists, waiver and insurance forms, documentation of divers' Certification cards, a list of safety rules and regulations, etc. If your cleanup is sponsored by a dive train-

ing agency, check to see if their standard waiver form is sufficient; you may have to add additional clauses or create a
form of your own.

VIII. ASSIGN TEAMS

Being part of a team stimulates fun and an upbeat competitive spirit. Teams are a great way to involve a whole family
and a way for people to meet new friends. Most important, a
team format makes cleaning up more efficient and effective.
One team member can hold the trash bag, another person
can record the trash which is found, while a third team member deposits the debris in the bag. You can offer prizes for the
team that collects the largest amount of garbage or the team
that finds the most "unusual" garbage. Try to give each team
its own distinctively colored hat or visor, shirt, name badge
or arm band, and make sure that team leaders are easily
identifiable.

IX. HAVE PROPER EQUIPMENT FOR THE CLEANUP

Cleanup participants should be advised to wear comfortable
clothes, sunblock, and hats. You should supply large trash
bags and disposable gloves for everyone handling debris. If

An effective cleanup must have correct equipment and a safe dive plan.

Photo: Scott Jones/Courtesy PADI–

Photo: CMC–

Cleanups can center around a specific problem in your local area.
Pictured: In Florida, volunteers picked up 304 miles of lost monofilament line in three hours.

possible, try to have recycling bins on hand. You must either have vehicles to transport the collected trash to a disposal facility or make arrangements with your local sanitation department.

For underwater cleanups, divers should have all of their standard scuba gear, including an alternate air source, buoyancy compensator, dive knife, as well as catch bags, wire cutters (if available), and buoy markers. Other helpful items to have available are ear wash and fresh water to clean off divers and their gear. Additionally, all divers should wear exposure suits and neoprene gloves for protection, even in warm water. **Equipment should always include an ample supply of rescue gear and first aid supplies.**

X. HAVE A SAFE DIVE PLAN

Even though an underwater cleanup can be fun, you should *never* deviate from or eliminate any element of safe diving. One or more supervisors should constantly oversee the entire cleanup. If you have many participants and must spread the cleanup out over a large area, try to have the leaders use walkie-talkies to maintain contact.

Cleanups can be fun and creative. Pictured: A unique Statue of Liberty, constructed by an artist from thousands of Tampon applicators which he collected during New Jersey coastal cleanups.

–Photo: CMC

Every participant should be given a written copy of your safety rules and a dive plan which specifies maximum allowable dive time, maximum depth, the boundaries of the cleanup area, and the correct ascent rate. The dive masters should be in charge of making equipment checks, buddy assignments, a review of communication signals, recording divers' Certification cards and credentials, and keeping a written log of dive teams going in and out of the water. Be sure that you hold a dive briefing to reinforce the dive plan and emphasize water conditions, search and recovery procedures, and safety protocol for handling debris. Extreme conservation precautions and buoyancy control should be emphasized in fragile environments, such as coral reefs, so that divers do not tear delicate soft coral or collide with coral or rub off its protective mucus coating.

Divers should not attempt to salvage dangerous or heavy objects. Questionable items should be marked with buoy markers, so that they can later be retrieved by professional salvage teams or local police divers. Land and water based support personnel should be on hand (wearing gloves) to retrieve debris from the divers. **Most important, a sufficient number of trained rescue personnel should be correctly positioned and equipped to deal with any underwater, surface, or land emergency.** Many local municipalities will be glad to lend assistance by supplying EMS, U.S. Coast Guard, police, and professional rescue dive team members.

XI. Use Data Identification Cards

One of the primary goals of a cleanup is helping to compile an official record of the debris that is found. The Center for Marine Conservation developed standardized data cards which list the items most often found at cleanups. These data cards are available from CMC at 1725 DeSales St. NW, Washington, DC 20036. Data cards are divided into eight major categories: plastic, glass, Styrofoam, rubber, metal, paper, wood, and cloth. Within each category there are listed "indicator items," to identify specific debris sources. CMC has identified 28 indicator items which fall into five categories:

1. fishing and boat gear;
2. galley wastes from ship kitchens;
3. operational wastes from vessels;
4. waste from offshore petroleum operations; and
5. medical waste.

Completed cards should be sent to the Center for Marine Conservation, where the information goes into an extensive database. Continuous data collection helps monitor and enforce the MARPOL Annex V Law. When illegal trash is identified, such as material from a particular cruise line, the offender is confronted and faces heavy fines. It is also possible for private persons to bring actions against violators.

XII. Make Your Cleanup Fun

Plan fun activities in conjunction with your cleanup. Free refreshments, a raffle with prizes, free souvenirs, such as bumper stickers, tee shirts, hats, Polaroid photos, etc., will add to the enjoyment of the event. With advance planning, cleanup contests are another option. In Florida, for example, where discarded fishing line is a particular problem, there are contests during cleanups to measure which team can haul in the most fishing line. In 1988 one team collected 304 miles of fishing line in three hours!

For more information about underwater clean-ups, see the PADI brochure, "How to Conduct an Underwater Cleanup," and PADI's new cleanup manual which contains a sample cleanup waiver, checklist, and other important information.

ECOTOURISM AND CONSERVATION ABROAD

"We cannot live only for ourselves. A thousand fibers connect us with our fellow man: and among those fibers, as sympathetic threads, our actions run as causes, and they come back to us as effects." — **Herman Melville**

According to the U.N.'s World Tourism Organization, there were 528.4 million international tourist arrivals in 1994. Their direct tourism expenditure was US $416 billion, this makes up 6% of world Gross Domestic Product. WTTC claim that tourism should generate more than US $3,000 billion (directly and indirectly) worldwide in Gross Annual Output by 1995, this accounts for 11.4% of all consumer spending and 11.3% of all capital investment. Cater (1994) indicates that "international tourist arrivals are set to doubled between 1990 and 2010 from 456 million to 937 million, the growth rate peaking at 4.4% in the year 2000. Within the tourist industry, ecotourism–travel based on the ecology of an area and that promotes conservation in that area–is one of the newest and fastest growing entities; its impact is being felt around the world. In a recent survey, more than eight million Americans said they had already taken an ecotour, and a further thirty-five million said they were likely to do so within the next three years.[1] The Ecotourism Society estimates that nature travel will grow by twenty to twenty-five percent this year alone. In only a few short years, a barrage of villages, islands, and countries have developed economic and conservation strategies incorporating ecotourism, and conferences and organizations on ecotourism are proliferating.

Whereas ecotourism has raised the promise of conservation and sustainable-development opportunities worldwide, to date there are no universal standards and guidelines for this new industry, nor are there strategic plans in the development stage which will protect the resources which are being exploited by tourism. There are roughly 1000 national parks in the world today, but few of them are in the undeveloped nations which need regulations most. Even in

– Photo: Mort and Alese Pechter

Ecotourism is a double-edged sword; it promises increased revenue and sustainable development, but it can over-stress fragile environments.

the developed countries, many resource protection plans are not in writing, and some parks have inadequate staff and funding. All of this has led to confusion, or to ad hoc creation of rules which may vary greatly from country to country and among tourist sites within the same country. This has also allowed ecotourism to be misused by individuals who just want to capitalize as fast as they can on a lucrative and "politically correct" trend.

Regardless of which definition one chooses, ecotourism is not just a simple process in which visitors "tread lightly," leaving minimal impact on environments they encounter. It is travel that actually *accomplishes* conservation and positively impacts local communities and natural resources in the process. Just hiking through a bird sanctuary or diving an exotic coral reef may be a wonderful adventure, but it is not "ecotourism." A minister of tourism in one African country recently declared that he was encouraging ecotourism, but when questioned further, could not explain how, why, or what the term meant.

In fact, some "ecotourism" can actually cause harm to fragile habitats and indigenous cultures, as witnessed by the uncontrolled tourism and poaching which marred Kenya's national parks in the late 1980s. According to the ecological principle of "carrying capacity," all protected areas are limited in the number of tourists that they can accommodate without altering and degrading their ecosystems (i.e., eroding soil, deforestation, creating sewage and litter, disrupting feeding and mating activities). If ecotourism is not responsibly controlled, the rapid increase will overload fragile areas, with disastrous results. A striking example of this problem can be seen in Nepal, as reported by the National Audubon Society. The number of tourists to Nepal increased fivefold from 40,000 in 1970 to 223,000 in 1986. Over the same period, the number of eco-tourists (mostly trekkers) tripled, from 12,600 to 33,600. This caused a virtually overnight emergence of more than 200 mountain lodges and the clearing of large areas in order to supply fuel wood for lodges and trekkers.

In an effort to lay down some guidelines for protecting fragile environments from over-exploitation by tourists, the Ecotourism Society, a non-profit group, has defined ecotourism as "responsible travel which conserves environments and sustains the well-being of local people." This includes everything from travel which produces minimum waste products to ensuring that the money spent by tourists goes back to and is used to enrich the area in which it was spent.

The National Audubon Society has established is own travel ethics for environmentally responsible behavior. These tenets include leaving wildlife and habitats undisturbed, disposing of waste correctly, strengthening the conservation ethic, enhancing the natural integrity of the places visited, prohibiting the sale and trade of endangered species, and respecting the sensibilities of local cultures.

Ecotourism can turn pristine habitats into bustling resorts for large numbers of vacationers.

Photo: Mort & Alese Pechter–

Another attempt to create a universal lexicon for ecotourism has been made by The Marine Conservation Society. MCS has published a "Visitor's Code," which is distributed by an international hotel chain in resorts in the Seychelles and Djibouti and by dive operators throughout the United Kingdom. The Ecotourism Society and the Audubon Society have also recently published handbooks for ecotourist guides, which offer advice on everything from disposing of trash in the wild to non-invasive techniques for photographing wildlife.

Although the ecotourism industry as a whole has yet to draft official standards and procedures, tourist boards, private travel agencies, tour operators, local guides, developers, even airlines, are jumping on the ecotourism bandwagon faster than you can say "rain forest." And while the scientific and environmental communities are debating the effects of ecotourism on protected habitats and species, local governments are exploring this fast-growing phenomenon with a mixture of fear and fascination.

One of the major complaints against many would-be ecotourist initiatives is that they do not involve local people in planning or implementation, leading to resentment and impoverishment. The World Bank estimates that only forty-five percent of money spent by tourists remains in Third World countries. Of the money spent by thousands of trekkers in Nepal every year, only twenty cents of every three tourist dollars goes to the local villages. Ecotourism can only be effective if it generates enough money and jobs for the natives in indigenous countries so that they are motivated to protect their environment. For example, if ecotravelers spend enough money to enjoy the coral reef areas in Southeast Asia for their aesthetic value, the locals will be motivated to stop mining coral for building materials. One of the most dramatic examples of how ecotourism can go hand-in-hand with conservation can be seen in Rwanda, where tourism is largely responsible for saving the nation's gorillas from extinction. The gorilla was threatened both by poachers and by local farmers who were destroying the gorilla's habitat for agriculture. By creating a wildlife preserve, the Parc des Volcans, Rwanda obtained an international tourist attraction and a major source of revenue.

MARINE PARKS

In marine and coastal environments, ecotourism has succeeded best through cooperative efforts of government, local people, and the tourism industry. An example is marine parks. Marine parks can now be found in many areas of the world, from Central America to the South Pacific to the Red Sea. In Bora Bora in French Polynesia, hotels have helped the government establish marine reserves on adjacent reefs. On Bonaire, a tropical haven in the Netherlands Antilles, a marine park was created to protect the beauty and health of the coral reefs which are visited by millions of divers a year. Visitors are charged a ten dollar yearly user fee, and before being allowed on the reefs, divers are required to participate in a reef etiquette and buoyancy control class. The user fee helps staff the marine park and pay for the maintenance of mooring buoys and research. Although locals initially feared that instituting a user fee would hurt tourism, the user fee is now strongly support-ed by the local dive shops and resorts, the Bonaire Tourist Board, as well as by the legislative and scientific communities.

Another marine park, on the small volcanic island of Saba, has created various zones for divers and fishers and the general public. Saba's marine park provides the mainstay of the island's revenue from tourism. The National Park and Biosphere Reserve in the U.S. Virgin Islands nets a profit of twenty-three million dollars a year.

A large marine park which has met with success is based in Canberra, Australia, on the Great Barrier Reef. The Great Barrier Reef Marine Park Authority is set up to respond to a number of interest groups and the environmental pressures to the reef that they represent. The Authority is advised by a team of consultants made up of representatives from groups with a vested interest in the reef and the surrounding ocean waters: fishermen, politicians, tour guides, aborigines, scientists, and conservationists. Special zones have been established to keep the various activities separate. The Authority's power to enforce the zoning uses even takes prece-dence over most state and federal legislation that might conflict with park goals.

On small islands, it may be the local dive guides, whose liveli-hood depends on the environment, who are in the forefront of con-servation. However, in highly populated marine and coastal resort areas, such as the Florida Keys, the challenges of ecotourism are more complex, such as controlling sewage treatment and litter dis-posal, and may necessitate the involvement of many different entities.

CHOOSING AN ECOTOURIST TRIP

Now, more than ever, there is a broad spectrum of ecotourist vacation/field expeditions which offer participants the opportunity to work with scientists, collect data on the status and environmental

–Photo: Paul Humann/REEF

*REEF is one of the organizations which conduct marine research
and conservation trips for divers.*

requirements of various marine species, monitor the effects of natural and human impacts, and help revitalize damaged habitats. The
fees from these trips also contribute directly to environmental protection by supporting conservation-directed research.

Anyone can and should participate in ecotourist field trips, and
very few programs require prior experience in research or marine
science. You should, however, obtain as much information as you
can beforehand and make sure that your trip meets your needs and
budget. Ecotourism can involve Spartan accommodations and
extreme physical challenges or be the extreme opposite–"sybaritic
travel," with luxurious trappings and high price tags.

There are currently more than 5,000 tour operators involved
with nature or adventure travel, not all of whom have proper experience or credentials. To ensure that your trip is a bonafide ecotourist experience check the outfitter's credentials thoroughly
beforehand and ask the following questions:

1. How much of the trip's revenue will stay in the local community?
2. Will you be staying in facilities and engaging in activities
   that minimize environmental damage?

3. How large are the tour groups and what is the ratio of guides to tourists?
4. Do the tour guides speak the language of the country you will be visiting?
5. Does the company donate a portion of ecotourist revenue to promote ecological research, to repair damage to the environment, and/or to develop parks and manage natural resources in the countries in which it conducts tours?
6. Does the company seek to reduce the environmental impact of tourists?
7. What form of transportation will you be using?
8. How will garbage be disposed?
9. Does the tour interact meaningfully with local culture and people?
10. Will photographing of wildlife be permitted, and if so, will it be monitored to minimize negative impacts on wildlife?

–Photo: Stephen Frink

Divers are needed to do surveys of marine habitats and their inhabitants.

Besides the duties expected of a good tour operator, each travel-er should also assume responsibility for "treading lightly" on the environment visited. The following are recommendations and a useful check list to make your eco trip successful and meaningful:

RESPONSIBILITIES OF THE ECOTRAVELER

1. Before embarking on an ecotour, you should learn as much as you can about the ecology and culture of your destination.
2. Always get a good on-site orientation.
3. It will not only familiarize you with the environment, but will help you protect it and appreciate it as inhabitants.
4. Make sure your diving skills are up to date and environmentally safe.
5. This may require a refresher course or practice in the pool first.
6. Even on land, follow basic rules. Enjoy but do not disturb wildlife or habitats.
7. Refrain from urging your guide or Divemaster to break the rules.
8. Choose small tour operators and groups with a good ratio of guides to tourists.
9. Avoid crowding
10. Interact with and show respect for the local people.
11. Respect local traditions and customs.
12. Obey all local laws.
13. Bring a minimum of disposable items and dispose of refuse correctly.
14. Participate in activities which protect and enrich the environment.
15. Spend your money on local crafts and cuisine.

ECOTOURIST CHECKLIST

a. check credentials of tour agency, resort, etc. in advance
b. proper equipment and clothing
c. immunizations shots
d. passport and visa

e. emergency medical kits
f. contingency plans
g. emergency phone numbers
h. guide and/or dictionary of local language
i. local currency
j. register camera and video equipment at customs
k. bring luggage that is rugged and lightweight
l. sun block, insect repellent, headgear

In this chapter, ecotravel groups and organizations are divided into two sections. The first section represents a sampling of trips specifically for undersea and scientific research. The second section contains a broader range of ecotravel experiences for divers as well as non-divers. On ecotourist research trips, participants do not need a prior scientific background and are usually trained on-site. However, if the trip includes scuba diving, you must show proof of scuba certification. Also, some dive training agencies, universities, and colleges award CEUS and academic credits to participants in these programs.

Individuals interested in more general information about ecotourism may also contact the Ecotourism Society (P.O. Box 755, North Bennington, VT 05257 (802) 447-2121). The Ecotourism Society is an international, non-profit organization dedicated to finding the resources and building the expertise to make tourism a viable tool for conservation and sustainable development. The Society documents the best techniques for implementing ecotourism principles by collaborating with a global network of professionals actively working in the field. The Ecotourism Society offers its members a consultation and referral service and a quarterly newsletter. *Ecotourism Guidelines,* a publication of the Society, is designed for nature tour operators.

If you are interested in keeping abreast of trends in ecotourism or want specific information about the cultures which you are going to visit, there are several ecotourist publications available:

Buzzworm Magazine Guide to Ecotravel, published by Buzzworm, Inc. 2305 Canyon Blvd., Suite 206, Boulder, CO 80302.

The Green Travel Sourcebook by Daniel Grotta and Sally Wiener Grotta, John Wiley & Sons, NY.

Cultural Survival, published quarterly by Cultural Survival, 215 First St., Cambridge, MA 02142

National Geographic Traveler, a bi-monthly magazine published by the National Geographic Society, 17th and M St., Washington, DC 20036.

FIELD TRIPS AND RESEARCH PROGRAMS FOR DIVERS

Bermuda Biological Station for Research

Diving enthusiasts of all levels and professionals who want to be on the cutting edge of marine science and work with the world's elite scientists can literally have a "field day" with the Bermuda Biological Station for Research's innovative new field programs. The Bermuda Biological Station for Research (BBSR), a U.S. non-profit organization, hosts some of the world's most renowned marine scientists and research teams and has created specialized environmental/conservation field programs specifically for recreational divers. BBSR staff and resident scientists will supervise small groups of divers during seven-day programs (ten-day and two-week programs can be arranged by special request). The programs include lectures, AV presentations, and in-water sessions in professional marine life surveys, assessment of natural and human-induced reef damage, problems related to island fisheries, use of Remotely Operated Vehicles (ROVs) for deep diving research, mariculture, and the new SeaWifs satellite for validating measurements and predicting global environmental changes.

NOTE: The Bermuda Biological Station for Research is one of only 12 designated SeaWifs satellite sites in the world. The SeaWif and the E.O.S. (Earth Observing System), which will be launched by NASA in 1998, circle the earth and use powerful sensors to take a global snapshot of the earth and its oceans. Incredibly, these satellites are able to measure actual wave height anywhere in the world, as well as the reflection and absorption of light by the water and a variety of other atmospheric and environmental factors. The BBSR receives the satellite data direct from NASA via a special computer link-up. It is hoped that by combining satellite data with information relayed to the Bio Station by several high-tech buoys throughout the Atlantic, scientists will be able to monitor the entire ocean, much as forecasters now do for the weather.

Additionally, BBSR is on the forefront of determining the ocean's role in global climactic change and is one of just two institutions in the world funded by the U.S. National Science Foundation to carry out research on determining long-term changes in the open ocean. The BBSR has also been selected to conduct the Joint Global Ocean Flux Study. The BBSR is currently pioneering methods and data for risk prediction in global climactic changes and developing technology by which scientists can predict earthquakes, El Ninos, greenhouse effect, coral bleaching, etc.

Divers who attend the BBSR programs, therefore, will be able to use state-of-the-art deep sea exploration equipment, speak with the scientists who are developing these breakthroughs, and have access to a database which is unavailable anywhere else. The BBSR library contains over 20,000 volumes, 150 current serial publications, DIALOG, OMSET, and TELENET computer access.

Activities for amateur and professional underwater photographers and videographers include training in high tech digitized video along a transect and special techniques for deep diving research.

BBSR field programs not only enhance a diver's knowledge of local reef ecology, oceanography, scientific technology, and Bermuda's history, but they focus on the most important and controversial conservation issues of our time–the importance of global biodiversity and the scientific, socio-economic, and political problems involved in the interactions between humans and the ocean.

BBSR offers seven-day field programs from August through October. Field program dives are done aboard BBSR research vessels and are supervised by BBSR staff and guest experts. Dives can be made on Bermuda's reefs and historic wrecks, such as the *Constellation* (the medical supply ship laden with cocaine vials, which was featured in the film *The Deep*). BBSR scientists also conduct blue water research activities, in which divers are taught to collect organisms in jars while diving in two to three miles of open ocean.

Bermuda Biological Station For Research
St. George's GE 01, Bermuda
(809) 297-1880

REEF (Reef Environmental Education Foundation)
REEF, a non-profit organization, was founded in 1991 by marine life photographer Paul Humann and publisher Ned DeLoach to provide an opportunity for recreational divers to become active monitors of the underwater wilderness. During their 25 years of underwater photography and research for *The Reef Set*–their comprehensive marine life identification guides to Florida, the Caribbean, and the Bahamas–Humann and DeLoach became disturbed by the marked decline of the reef system's vitality and by the lack of information about the marine environment compared to terrestrial ecosystems.

By setting into motion long-term education and research programs, REEF seeks to unite the scientific and recreational diving communities to facilitate this important task. REEF has begun by offering five-day reef survey programs for divers of all levels; no prior background in marine biology is necessary. These field survey programs are conducted in popular dive destinations in the U.S. and the Caribbean. A REEF trip is an excellent learning vacation coupled with a lot of diving and a chance to work with REEF personnel, scientists, and other divers interested in the reef's natural history.

After completing a REEF survey program, participants will be able to quickly distinguish between the major families and identify nearly fifty of the most commonly sighted fish species on a reef dive. Besides marine life identification, divers are taught to gather species-sighting data and enter the information on bubble sheets, which are later catalogued in a central computer bank. As the skills of volunteer researchers improve, more sophisticated data can be collected, and sighting information will expand to include corals and other marine invertebrates. From these efforts, scientists and environmental groups will, for the first time, have a working knowledge of the region's wildlife distribution, population trends, and behavior. This information will, in turn, help monitor the health and biological diversity of major reef areas, making REEF field trips valuable ecotourist experiences.

REEF
PO Box 246
Key Largo, FL 33037
(305) 451-0312

American Littoral Society Field Trips

The American Littoral Society, a national, non-profit public interest organization of professional and amateur naturalists, was founded in 1961 at Sandy Hook, New Jersey by a small group of divers, naturalists, and fishermen. Their goal was to encourage a better understanding of aquatic environments and provide a unified voice advocating conservation. The shore and adjacent wetlands, bays, rivers–*the littoral zone*–is the special area of interest and concern for members of the American Littoral Society. In addition to sponsoring conservation legislation, ALS is currently conducting the world's largest volunteer tag-and-release program, producing solid information on migration and growth rates of marine fishes.

The American Littoral Society is an active group which devotes much of its energy to outdoor field experiences–hikes, dives, and explorations of the coastal strand. Throughout the year, the Society offers a large variety of trips to wilderness habitats, led by experienced naturalists and conservationists. A jungle trek in the Galapagos, eight days of hiking, stream walking, and shoreline exploration around Alaska's rocky coasts, and canoeing/camping trips which explore the natural history of the Okefenokee Swamp are just a few examples.

A popular one-day American Littoral Society annual event is the shark and tuna tagging expedition off Point Pleasant, New Jersey. Participants are taken on small, fast boats and guided by local experts to find, hook, fight, tag, and release mako, blue and dusky sharks, and tuna fish.

American Littoral Society's seven-day diving trips focus on marine conservation. Trips are available in dive locations such as Puerto Rico, Cozumel, Galapagos, and Belize, and are led by environmental executives, such as Alexander Stone, the Director of Project Reefkeeper, as well as local experts. These trips include ecology lectures and workshops, meetings with local conservation leaders, shore walks, tours of marine parks and estuarine lagoons, a visit to the coral reef natural history museum (in Cozumel), water quality monitoring, and other field projects. Some of the Littoral Society's recent expeditions included a Cape Cod Whale Watch, Ecology Week in Alaska, Hawaii Ecology Dive Week, and a Chesapeake Bay Oyster/Fossil Dive.

American Littoral Society
Highlands, NJ 07732
(908) 291-0055

Oceanic Society Expeditions

Oceanic Society Expeditions (OSE), an affiliate of the environmental advocacy group Friends of the Earth, is a non-profit organization, completely self-supporting and independent of outside funding and corporate grants. OSE promotes environmental stewardship, education, and research through ecotourism. Oceanic Society Expeditions is seeking certified scuba divers to help carry out long-term marine biodiversity and coral reef monitoring. The research results are used to provide baseline data needed to better protect ecosystems and habitats through intelligently drafted legislation.

OSE trips are organized and conducted in conjunction with the OSE scientific advisors, which include such notables as NOAA scientist Dr. Sylvia Earle and Stephen Leatherwood, Chairman of the World Conservation Union/IUCN Cetacean Specialist Group. Research specialty certification is available on most expeditions.

OSE trips often involve diving, but many are open to non-divers too. OSE conducts dolphin research and surveys in the Bahamas, Honduras, and the Amazon (where you can swim with pink dolphins and explore the rain forest). Other OSE projects are a Monterey Bay biodiversity study and a surface and underwater

research survey of dolphins in Belize. Team activities in a sea turtle and manatee project in Belize include surveying a lagoon to map the distribution of the manatees, locating nesting turtles, collecting biological data, chaperoning hatchlings to the sea, and building and repairing turtle nesting cages that protect the eggs from predators.
Oceanic Society Expeditions
Fort Mason Center, Bldg. E
San Francisco, CA 94123
(800) 326-7491

Oceanographic Expeditions
Oceanographic Expeditions is a company that works with Dive Leaders and dive store owners who want to offer their students and customers a marine science/conservation-oriented dive vacation. Oceanographic Expeditions customizes group trips to Cozumel, Roatan, the Dry Tortugas, and the Flower Garden Banks. During these expeditions, divers are taught by oceanographers how to collect oceanographic data and how oceanographic conditions affect marine life, and they also can assist ongoing research.

"What is the ocean's maximum underwater visibility?" "Do biolumi-nescent plankton glow in daylight, and what do they look like under a microscope?" Answers to these and a myriad of other questions are explained in the lectures. The academic portions of the program are easy to understand, and participants can also learn to take repetitive photographs, time lapse video, and count fish in an easy, hands-on context. The data collected is used by the Florida Marine Research Institute, the Florida Institute of Oceanography, the Bay Islands Conservation Association, Texas A & M University, as well as the Flower Garden Banks and Florida Keys National Marine Sanctuaries.
Oceanographic Expeditions, Inc.
4418 Saint Ann Street New Orleans, LA 70119
(504) 488-1572

NOAA'S "Great American Fish Count"
This new program allows volunteer sport divers to help NOAA's scientists establish an ongoing marine life databank to help sanctuary managers develop resource management policy. NOAA's program began in June 1994 at many of the National Marine Sanctuaries.
NOAA National Marine Sanctuary Program Office
1305 East West Highway, Bldg. 4
Silver Springs, MD 20910
(301) 713-3125

CEDAM International

CEDAM offers educational and conservation-oriented trips for divers in many premiere diving destinations, such as the Red Sea, Palau, the Caribbean, Mexico, Bonaire, the Cayman Islands, and Belize. Trips are led by well-known marine scientists, archeologists, explorers, and photographers. Volunteers participate in underwater surveys, help collect live fish for display at public aquariums, and are included in other conservation activities. CEDAM is affiliated with the National Geographic Society, the New York Aquarium, the Center for Marine Conservation, and Wildlife Conservation International.

CEDAM International
One Fox Road
Croton-on-Hudson, NY 10520
(914) 271-5365

University Research Expeditions Program

Participants in these trips to worldwide locations are led by scientists researching crucial global environmental problems. Tourists and University of California scholars conduct scientific research with an emphasis on the use of scientific methods to monitor ecosystems. University Research Expeditions also conducts many programs for non-divers. Trips to the Andes of northern Ecuador, for example, pair land-based travelers and scientists to inventory the orchids, bromeliads, and other flowering plants of the threatened cloud forest.

University Research Expeditions Program
University of California
Berkeley, CA 94720
(510) 642-6586

ECOTOURIST ADVENTURE AND WILDERNESS TRIPS

The following organizations and agencies offer a variety of ecotourist trips for water sports enthusiasts, including divers and snorkelers, as well as non-water related adventures.

Action Whitewater Adventures

Whitewater rafting trips in Idaho, Colorado, Grand Canyon, and California.

PO Box 1634
Provo, UT 84603
(800) 453-1482

Adventure Center

An international ecotourist organization offering trips in over 100 countries, including Africa, the Middle East, Asia, Latin America, the South Pacific, and Europe. Specializes in interaction with local cultures and low environmental impact activities.
1311 63rd St., Suite 200
Emeryville, CA 94608
(510) 654-1879

Alaska Discovery, Inc.

Low-impact exploration via sea-kayaking, canoeing, and rafting.
234 Gold St.
Juneau, AK 99601
(907) 586-1911

Alaska Wildland Adventures

Specializes in Alaska natural history safaris and authentic wilderness expeditions.
PO Box 389
Girdwood, AK 99587
(907) 783-2928 (800) 334-8730

Amazonia Expeditions

Conducts ecologically-oriented expeditions in the Amazon rain forest.
1824 NW 102nd Way
Gainsville, FL 32606
(904) 332-4051

American Wilderness Experience, Inc. (AWE)

This organization specializes in ecologically sensitive travel for small groups in prime adventure areas of the world. Destinations include U.S., Canada, Mexico, Virgin Islands, Australia, New Zealand, Peru, and Belize. AWE activities including snorkeling and diving, canoeing, sea kayaking, horse packing, llama trekking, whitewater rafting, mountain biking, and natural history treks.
PO Box 1486
Boulder, CO 80306
(303) 444-2622 (800) 444-0099

Baikal Reflections, Inc.

Offers environmental and cultural interactive trips to Lake Baikal and other parts of the Commonwealth of Independent States. Activities include ocean kayaking, backpacking, cycling, sailing, and diving.

13 Ridge Rd.
Fairfax, CA 94930
(415) 455-0155 (800) 927-2797

Baja Expeditions
Offers environmental and natural history expeditions in Baja,
Mexico and Costa Rica. Adventures include whale watching, scuba
diving, sea kayaking, mountain biking, and wilderness sailing.
2625 Garnet Ave.
San Diego, CA 92109
(619) 581-3311 (800) 843-6967

Bill Dvorak Kayak & Rafting Expeditions, Inc.
This company focuses on trips that support river conservation and
low-impact environmental activities. Whitewater rafting, kayaking,
canoeing, horseback riding, mountain biking and instructional seminars are offered in the U.S., Mexico, New Zealand, and Australia.
17921-Z U.S. Hwy. 285
Nathrop, CO 81236
(719) 539-6851 (800) 824-3796

Canadian River Expeditions
Led by naturalists and environmental experts, CRE arranges
wilderness expeditions to remote regions in western Canada and
Alaska. Trips focus on understanding and conservation of river and
coastal ecosystems.
3524 W. 16th Ave.
Vancouver, BC V6R 3C1, Canada
(604) 738-4449

Earthwatch
A major promoter of scientific research expeditions throughout the
world. Under the guidance of experts, trips focus on environmental
education, monitoring global change, conservation of endangered
species and habitats, cultural exploration, and promoting world
health and international cooperation.
680 Mt. Auburn St., Box 403
Watertown, MA 02272
(617) 926-8200

Fantasy Adventures of Earth, Inc.
Wilderness adventures include scuba diving, horseback riding,
mountain biking, mountain treks, whitewater challenges, and sail-

ing in the U.S., Canada, Peru, New Zealand, Belize, Alaska, and Hawaii.
PO Box 368
Lincolndale, NY 10540
(914) 248-5107

Friends of the Earth Ecotours
A non-profit environmental organization which donates all proceeds from its trips to preserve the areas visited and to the preservation of Japan's forests and wetlands. Professional naturalists lead wildlife tours in Japan, Hawaii, Siberia, Vietnam, and Antarctica, with an emphasis on understanding and appreciating natural ecosystems.
4-8 15 Naka Meguro
Meguro-ku, Tokyo 153, Japan
(81-3) 3760-3644

International Expeditions
Global adventure travel, ecotours, and a series of rain forest workshops open to environmentalists, scientists, and travelers. The first workshop in 1991 raised $60,000 for the Amazon Center for Environmental Education and Research, a base of study for scientists and tourists.
One Environs Park
Helena, AL 35080
(800) 633-4734

International Oceanographic Foundation
An organization whose trips include whale watching, scuba diving, natural history, sailing, and cultural studies. Special emphasis is on oceanography, study of cultures of coastal cities, marine mammals, and coral reef ecology. Trips locations include Baja, California, Alaska, Arctic regions, Amazon, Costa Rica, South Africa, Florida coast, Mississippi, and Egypt.
4600 Rickenbacker Causeway
Miami, FL 33149
(305) 361-4697

Island Packers
Research oriented trips to the Channel Islands, with an emphasis on preservation of terrestrial and marine wildlife.
1867 Spinnaker Dr.
Ventura Harbor, CA 93001
(805) 642-1393

Journeys International
4011 Jackson Road
Ann Arbor, MI 48103
(800) 255-8735

National Audubon Society
NAS conducts a variety of ecotourist trips around the world, which are led by naturalists and local experts. Ecotravel options include cruises in the Caribbean on clipper ships.
700 Broadway
New York, NY 10003
(212) 979-3066

The Nature Conservancy
1815 North Lynn Street
Arlington, VA 22209
(703) 841-5300

Outer Edge Expeditions
This outfit leads small groups on close encounters with wildlife in Australia (Great Barrier Reef), the Amazon jungle, Canada, Peru, and other exotic locations, with an emphasis on ecology and conservation. Activities include kayaking among whales, swimming with pink dolphins, traveling in a dugout, camel trekking, and dog sledding.
45500 Pontiac Trail, Suite B
Wallod Lake, MI 48390
(313) 624-5140 (800) 322-5235

Overseas Adventure Travel
349 Broadway
Cambridge, MA 02139
(800) 221-0814

PADI Travel Network
A division of PADI, the largest dive training agency, this is an extensive network which connects divers to numerous resorts around the world, many of which are dedicated to environmental conservation.
1252 East Dyer Rd., Suite 100
Santa Ana, CA 92705-5605
(714) 540-7234

Safari Consultants, Ltd.

Ecotourist trips throughout Africa (Botswana, South Africa, Zimbabwe, Kenya, Zambia, Tanzania) which give participants first hand exposure to learning about the effects of tourism, game management, and wildlife preservation under the direction of government and public institutions.
4N211 Locust Ave.
West Chicago, IL 60185
(708) 293-9288 (800) 762-4027

The Sierra Club

The oldest environmental vacation organization offers a large number of ecotourist trips as well as "service" trips each year through its Outings Program. In these programs travelers work with naturalists to repair damage to fragile ecosystems and maintain nature trails.
730 Polk Street
San Francisco, CA 94109
(415) 776-2211

Smithsonian Research Expedition Programs

490 L'Enfant Plaza, SW, Suite 4210
Washington, DC 20560
(202) 287-3210

Special Expeditions, Inc.

Expert naturalists, geologists, historians, and archeologists lead participants on environmental travel in exotic locations (Alaska, Baja, California, Columbia and Snake Rivers, U.S. Southwest, Caribbean, Belize, Arctic Norway, Galapagos, British Isles, Russia, Baltics, Egypt, Papua, New Guinea, Morocco, and East Africa). In some locations, customized trips are available aboard expedition ships or on tented safaris.
750 Fifth Avenue
New York, NY 10019
(212) 765-7740 (800) 762-0003

Victor Emanuel Nature Tours

PO Box 3308
Austin, TX 78764
(800) 328-8368

Wilderness Travel
801 Allston Way
Berkeley, CA 94710
(800) 247-6700

Wildland Adventures
Conducts worldwide trips which support conservation and community development projects in conjunction with its own affiliated non-profit conservation organization. Activities include wildlife safaris, trekking, jungle expeditions, ecology expeditions, personal cross-cultural interactions. Locations: Alaska, Andes, Amazon, Africa, Belize, Costa Rica, Himalayas, Mexico, and Turkey.
3516 NE 155th St.
Seattle, WA 96155
(206) 365-0686 (800) 345-4453

World Wildlife Fund
WWF's mission is the conservation of nature. To enhance members' understanding of natural systems and conservation challenges, WWF offers trips escorted by naturalists and wildlife photographers to areas of the world rich in wildlife. Destinations include Bonaire, Amazon River, Indonesia, the Weddell Sea, Chile, whale watching in Baja, California, Tanzania, Kenya, Seychelles, Madagascar, Costa Rica, Panama, Borneo, and Papua New Guinea.
WWF Travel Program
1250 Twenty-fourth St., NW
Washington, DC 20037
(202) 293-4800

Zegrahm Expeditions
Experts lead small groups on cultural, ecology, scuba diving, and photography expeditions to the Amazon, Antarctica, Arctic, Asmat, Botswana, Cambodia, Galapagos, and Vietnam.
1414 Dexter Ave. N, Suite 327
Seattle, WA 98109
(206) 285-4000

NOTES

CITIZEN ACTION

"Never doubt that a small group of thoughtful, committed citizens can change the world. Indeed, it is the only thing that ever has." — ***Margaret Mead***

Our commitment to the health of our water planet cannot stop at the water's edge. It is important that as many people as possible become actively involved in the decision-making process for environmental issues. Environmental problems are, in essence, people problems. The problems are largely caused by people and improvement can only evolve if people demand it. We must voice our opinions and concern to local and federal government officials about pollution, coastal management, sewage treatment, zoning laws, fishing regulations, and other relevant issues.

To be effective, you must learn where to direct your comments and involvement. Learning effective citizen action can be divided into two main categories: learning how to organize and express your concerns and learning how to make your opinions reach the correct authorities and agencies. This chapter, therefore, is a guide to citizen action, and will explain the basics of organizing your ideas and relaying them to local and government representatives.

No one person should expect to successfully tackle all of the global environmental problems at once, or all alone. The late Robert Rodale believed that "the only way to save the world is to save three lives." In other words, start small. You can experience the uniquely satisfying feeling of contributing to the improvement of the marine environment in just one area.

If you find yourself most interested in one particular issue, for example, marine mammal protection, become as knowledgeable about that issue as possible, and recruit others to join with you. You can be sure that others share your concerns and are willing to support your efforts.

In recent Roper Organization polls, for example, it was revealed that sixty-four percent of American adults interviewed said environmental protection was more important than economic gain, and four out of five Americans feel that the nation should "make a major effort to improve the quality of the environment." It is likely, therefore, that individuals and organizations in your area may already be working towards solutions to your specific concerns. Even the smallest contribution you make is important because it adds to the work being done by other people, and combined efforts produce success.

Do not be discouraged by preconceived and false notions of influence peddling, highly paid lobbyists, and perplexing decisions by government beyond your control. The importance of citizen action cannot be overemphasized! While an individual may not have the resources available to large organizations, individual citizens do have one powerful weapon–the vote. Government decision makers understand this, but they will not know what decision you think they should make unless you tell them. Time and time again, citizens who have made their voices heard have affected government decisions.

Furthermore, environmental issues are seldom won or lost definitively. Times change, lawmakers change, the environment and its problems change. Nothing is written in stone forever. The following appendixes will offer some tools for citizens to become more actively and effectively involved in marine conservation. The more we use these tools, the more effective the public's voice will be and the more our presence accepted when laws are enacted.

How to Organize and Express Your Concerns

1. **Stay informed.** Obtain as much information as you can from environmental organizations. Ask that you be placed on the mailing list of all local environmental organizations in your area. You should receive newsletters, updates, and announcements of upcoming public meetings. Find out when councils are being held and when particular issues will be open for public discussion

2. **Read the *Federal Register*.** The *Federal Register* is the formal notification document for federal activities in all agencies. (See Figure 19.) It is published daily, to update the public and government officials. You can obtain copies of the *Federal Register* at most public libraries. *Federal Register* notices are particularly helpful in providing specific names of people to

contact for more information on an issue or to whom to send your written comments about an issue. If your local library does not have a subscription to the *Federal Register*, encourage them to get one. The *Federal Register* is printed by the Government Printing Office in Washington, DC.

3. **Attend public meetings in your area and talk to local officials one-on-one.**
4. **Publicly voice your views at council hearings.**
5. **Write articles and letters to the editor of newspapers and magazines.**
6. **Contact your Congressmen and/or Senators.** Surprisingly few people in America (less than ten percent) ever write their elected officials, although we pride ourselves on the freedom to do so. Elected officials are the people whose votes will determine what price we will pay for the acts of government, either in dollars, in human lives, or in the quality of our environment. Call and meet with your U.S. Congress Representatives and/or Senators or their staff, either in their home offices or in Washington, DC. (See chapter 19)
7. **Learn how to lobby.**
8. **Join one or more environmental groups.**
9. **Start an environmental group or chapter in your community.**

This notice provides only the FIPS publication number, title, and the technical specifications number for each of the twelve standards being withdrawn:

-FIPS 3-1. Recorded Magnetic Tape for Information Interchange (800 CPI. NRZI)(ANSI X3.22-1973).

-FIPS 25. Recorded Magnetic Tape for Information Interchange (1600 CPI. Phase Encoded)(ANSI X3.39-1973).

-FIPS 50. Recorded Magnetic Tape for Information Interchange. 6250 cpi (246 cpmm). Group Coded Recording (ANSI X3.54-1976).

-FIPS 51. Magnetic Tape Cassettes for Information Interchange (3.610 mm (0.150 in) Tape at 32 bpmm (800 bpi). PE (ANSI X3.48-1977).

-FIPS 52. Recorded Magnetic Tape Cartridge for Information Interchange. 4-Track 6.30 mm (1/4 in). 63 bpmm (1600 bpi). Phase Encoded (ANSI X3.58-1977).

-FIPS 79. Magnetic Tape Labels and File Structure for Information Interchange (ANSI X3.27-1978).

-FIPS 93. Parallel Recorded Magnetic Tape Cartridge for Information Interchange. 4-Track. 6.30 mm (1/4 in). 63 bpmm (1600 bpi). Phase Encoded (ANSI X3.72-1981/R1987).

-FIPS 114. 200 mm (8 in Flexible Disk Cartridge Track Format Using Two Frequency Recording at 6631 bprad on One Side-1.9 tpmm (48 tpi) for Information Interchange (ISO 5654/2-1985).

-FIPS 115. 200mm (8 in) Flexible Disk Cartridge Track Format Using Modified Frequency Modulation Recording at (13262 bprad on Two Sides-1.9 tpmm (48 tpi) for Information (ISO 7065/2-1985).

-FIPS 116. 130 mm (5.25 in) Flexible Disk Cartridge Track Format Using Two-Frequency Recording at 3979 bprad on One Side-1.9 tpmm (48 tpi) for Information Interchange (ISO 6596/2-1985).

-FIPS 117. 130 mm (5.25 in) Flexible Disk Cartridge Track Format Using Modified Frequency Modulation Recording at 7958 bprad on Two Sides-1.9 tpmm (48 tpi) for Information Interchange (ISO 7487/3-1984).

-FIPS 118. Flexible Disk Cartridge Labeling and File Structure for Information Interchange (ISO 7665-1983).

Effective Date: This withdrawal is effective March 10, 1992.

For Further Information Contact: Ms. Shirley Radack, National Institute of Standards and Technology. Gaithersburg, MD 20899, telephone (301) 975-2833.

Authority: Federal Information Processing Standards Publications (FIPS PUBS) are issued by the National Institute of Standards and Technology after approval by the Secretary of Commerce pursuant to section Administrative Services Act of 1949 as amended by the Computer Security Act of 1967. Public Law 100-235.

Dated: February 4, 1992.

John W. Lyons,

Director.

(FR Doc. 92-5524 Filed 3-9-92; 8:45 am)

National Oceanic and Atmospheric Administration

Florida Keys National Marine Sanctuary Advisory Council; Open Meeting

Agency: The Office of Ocean and Coastal Resource Management. National Oceanic and Atmospheric Administration. Commerce.

Action: Notice.

Summary: The Council was established in December 1991 to advise and assist the Secretary of Commerce in the development and implementation of the comprehensive management plan for the Florida Keys National Marine Sanctuary.

Time and Place:

March 24 and 25, 1992 from 9 a.m. until adjournment. The meeting will take place at Buccaneer Resort, 2600 Overseas Highway, in Marathon, Florida.

Agenda:

1. Review goals and objectives for the Florida Keys National Marine Sanctuary.

2. Elect chairperson and vice chairperson and adopt by-laws for council.

3. Review and discuss possible water use zoning categories and criteria.

Public Participation: The meeting will be open to public participation and the last thirty minutes will be set aside for oral comments and questions. Seats will be set aside for the public and the media. Seats will be available on a first-come first-served basis.

Federal Domestic Assistance Catalog Number 11.429 Marine Sanctuary Program

Dated: March 5, 1992.

John J. Carey,

Deputy Assistant Administrator for Ocean Services and Coastal Zone Management.

(FR Doc. 92-5632 Filed 3-9-92; 8:45 am)

National Oceanic and Atmospheric Administration.

Modification of Scientific Research Permit.

Agency: National Marine Fisheries Service.

Action: Modification of Scientific Research Permit No. 648 (P #45D)

Notice is hereby given that pursuant to the provisions of the Endangered Species Act of 1973 (16 U.S.C. 1531-1543), the National Marine Fisheries Service regulations governing endangered species permits (50 CFR parts 217-222), and the conditions hereinafter set out, Scientific Research Permit No. 648, issued to the United States Fish and Wildlife Service, 75 Spring Streets, SW., Atlanta, GA 30303, on October 4, 1988, has been modified to authorize the taking of short-nosed sturgeon from waters in Georgia as well as from South Carolina. Eggs from up to twenty of the originally authorized fish may be used for studies of dioxin and furan loads.

This modification becomes effective upon publication in the Federal Register.

Documents pertaining to this Modification and Permit are available for review in the following offices by appointment:

Office of Protected Resources, National Marine Fisheries Service, 1335 East-West Highway, Silver Spring, MD 20910; and

Director, Southeast Region, National Marine Fisheries Service, 9450 Koger Blvd., St. Petersburg, FL 33702 (813/893-3141).

Dated: March 2, 1992.

Charles Karnella,

Acting Director, Office of Protected Resources, National Marine Fisheries Service.

(FR Doc. 92-5464 Filed 3-9-92; 8:45 am)

Department of Defense

Public Information Collection Requirements Submitted to OMB for Review

Action: Notice.

The Department of Defense has submitted to OMB for clearance the following proposal for collection of information under the provisions of the Paperwork Reduction Act (44 U.S.C. chapter 35).

Sample page from __Federal Register,__ Announcing a public meeting concerning the management of the Florida Keys National Marine Sanctuary.

GETTING INVOLVED IN SPECIFIC ENVIRONMENTAL ISSUES

"Here in the U.S., we must seize this opportunity to invest in our future by protecting our environment. If we fail to act, we lose. We lose jobs, we lose opportunities, we lose the ability to remain competitive in the international marketplace and we lose our natural resources."
  – Vice President Al Gore

As population and developmental pressures increase, there has never been a more critical time for the public to get involved in strategies to protect the marine environment. Regardless of where you live, there are issues which affect you and need your support.

Each of us can create an environmental success story. They happen as long as people care about their environment and get involved. After you have made a commitment to promote conservation, define your goals and keep your focus specific. Start by choosing an issue in which you want to get involved and list exactly what your concerns are and what changes you want made in that area. For example, if you live near a river, lake, or estuary, are you dissatisfied with the water quality? Does it meet federal standards? What pollutants are in the water and what are the sources of that pollution? Is the fish safe to eat? Is the water safe to swim in? What steps have been taken by local/state/federal agencies? Is there a local politician who has expressed particular interest in this problem? Is there an NGO (non-governmental environmental group) working on the issue? (See listing of NGOs in Appendix 3.) If so, be sure to contact these people and offer to pool your resources. There are several kinds of environmental groups you can join, depending on your interests and concerns. The options include local community groups (sponsored by schools, work and trade associations, church groups, or recreational centers), single-issue groups, or multi-issue groups such as the Audubon Society.

COASTAL ZONE MANAGEMENT

The burden of managing coastal development and conserving coastal areas falls upon the states for the most part, so your involvement begins at the state and local level. To become actively involved in decisions affecting coastal areas, contact local and state coastal program agencies. Becoming familiar with provisions of your state's coastal zone program will enable you to assess particular projects in your area. Getting involved in local zoning and permit decisions is particularly important.

Periodically, Congress revises the federal coastal zone management law and other equally important laws, such as the Clean Water Act. The Coastal Zone Management Act was reauthorized and strengthened by Congress in 1990, but every year Congress makes decisions on funding coastal zone agencies and major development projects. Get involved with local or national environmental groups that follow this issue and voice your opinion and concerns when you can.

COASTAL BARRIER PROTECTION

The future of many coastal habitats, including marshes, bays, estuaries, mangroves, etc., depends upon insuring that coastal barriers can continue to function. Congress has extended some protection to some coastal barriers through the denial of federal subsidies. But even in the absence of these subsidies, imprudent and environmentally damaging development can occur.

The key to insuring that coastal barriers are maintained is public scrutiny of permits for activities that damage coastal barriers. Potentially, the following activities, which often require permits from state or federal agencies, may have damaging effects: dredging and filling for construction of homes or other buildings and marinas; construction of seawalls, jetties, and riprap; construction of buildings in primary dunes or in marshes; construction of roads, causeways, and bridges; and extension of water and sewage lines to undeveloped areas. Coastal zone management agencies are generally required to provide an opportunity for public comment on permits for such activities. Contact your state coastal management agency for information on how you can be notified of pending permits.

Sewage and Other Discharges

The Environmental Protection Agency periodically issues proposed changes to existing regulations on point source pollution or proposes new ones. Public notice must be given when proposing or changing these regulations, and sometimes public hearings are held. You have the opportunity to comment on these new or revised regulations, either in person at a hearing or through written comments. You can also present your views on proposed National Pollutant Discharge Elimination System permits, again, usually in writing, and sometimes at public hearings. Contact the EPA regional office that covers your state. (See Appendix 3.)

At the local level, you can become involved with zoning and permit decisions. You can contact your local planning department or state coastal zone management agency to find out about proposed developments and industrial facilities planned for your area. They often hold public hearings on these matters and you can appear at these hearings and present your views. Also, the EPA is developing new approaches to controlling discharges of storm water and of sewage overflows from combined storm and septic systems. You can become involved by commenting on local plans to meet these new requirements (contact the appropriate state agency or the regional office of the EPA).

Urban and Agricultural Runoff

This is an area where you can take some direct action by yourself and make a difference. Things like used motor oil, paint, oven cleaners, brake or transmission fluid, antifreeze, rug and upholstery cleaners, pesticides, and furniture strippers are hazardous wastes. Try to reduce the amount of these products that you use. When you do use these products, dispose of wastes properly. Some of these products can be recycled–check with your local government or civic groups to find recycling programs. Some local governments accept some of these products for incineration, so call them to see what you can hand over to them. Call your local wastewater treatment plant to get information on disposing of liquid wastes.

Another important thing you can do in this area is to become involved with local zoning and permitting decisions. You can be involved at two levels: one, at the level of individual permit applications, where you can promote the inclusion of specific measures to limit non-point source pollution; and two, at the policy level where you can promote the development of policies, such as requiring new developments to have sewer systems rather than septic tanks. Contact your county or municipal government for more information.

You can also get involved in your state's coastal zone management program, as described above. This will give you the opportunity to contribute to decisions, such as the development of silting policies, at the state level that may affect non-point source pollution. In addition, you can monitor Congressional action relating to the Clean Water Act and the Coastal Zone Management Act and advise your Congressional representatives of your views on any revisions to these acts. Contact local or national organizations that work on these issues and find out how you can get involved. (See Appendices 2 and 3.)

TRANSPORT OF OIL AND HAZARDOUS SUBSTANCES

The federal government has the primary responsibility for promoting vessel safety. The Senate has the power to ratify international agreements on a variety of issues, from training standards to staffing to vessel design. You can get involved in that process. Contact your Senators for more information. (See Appendix 2.) Periodically the Coast Guard issues proposed safety regulations and gives the public an opportunity to comment on them and suggest changes and improvements. You can watch for these regulations and comment on them, usually in writing, and sometimes at public hearings. State governments control the requirements for the use of pilots by ships arriving from foreign ports. Many states are re-examining their requirements in this regard, as well as other matters affecting maritime safety. To find out what your state is doing about these issues, contact your state department of transportation or public safety or local port authorities.

OIL AND GAS DRILLING

The outer continental shelf oil and gas program is run by the Minerals Management Service, an arm of the Department of the Interior. They periodically prepare a nationwide five-year plan for offshore oil and gas development. In 1991, the Service published its proposed plan for oil and gas releasing from 1992 through 1997. Although the comment period on the overall plan is concluded, the specific lease sales will be announced in the *Federal Register*, asking for input from interested parties. You can suggest things for them to consider in the sale, and you can further suggest concerns to be addressed in the environmental impact statement they prepare on the sale. You can also comment on each Draft Environmental Impact Statement when it is released. Contact the Minerals Management Service (see Appendix 2) to learn the current status of the planning process. The Environmental Protection Agency, when preparing to issue an NPDES permit, also allows for public comment. Contact EPA Region IV or VI for information on the comment process.

The individual states have considerable power over the sale of drilling rights because the sale must be consistent with the state's coastal zone management plan. The state may block any sale that is found to be inconsistent with the state's plan. Citizens concerned about drilling rights sales in their areas, then, should also contact their state's coastal zone management agencies or the state agency charged with overseeing offshore oil and gas development.

WETLAND LOSSES

Enacted in 1990, the Coastal Wetlands Planning, Protection and Restoration Act established a federal task force and three programs to protect U.S. wetlands. The federal government has also been active in fighting wetland loss. Section 404 of the Clean Water Act regulates the discharge of dredged or fill material into U.S. waters and wetlands by requiring an environmental review and a permit issued by the Army Corps of Engineers. Although the review process is complex and involves several government agencies, it also allows for public input. When the Corps of Engineers receives a request for a permit, they generally prepare an environmental assessment. In some cases they will also prepare an environmental impact statement. You can comment on the scope of the EIS and on the contents of the draft EIS. There may be public hearings on the permit and the EIS and you can present your views there. You can also submit your views and comments in writing. To receive notices of proposed activities needing permits, contact the U.S. Army Corps of Engineers district nearest you. (See Appendix 2.)

Closer to home, you can contact the local and state coastal agencies in your locality. They have substantial input into decision making on coastal development, including that in wetlands. They can advise you of opportunities for you to assess projects in your area. You can also get involved in local zoning and planning activities. Things to be particularly concerned about include residential development in wetlands, disposal of dredged material in wetlands, and construction of canals through wetlands.

FISHERIES MANAGEMENT

The complexities of fishery management science have deterred interested conservationists from becoming involved in critical decisions affecting fisheries, endangered and threatened species, fisheries habitat, and the impact of industrialized fishing on marine habitat. As a result, fishery management decisions generally reflect the opinions of only those with a vested interest in catching fish. There are a number of opportunities for citizens to become involved in the fishery management process.

Under the Magnuson Fishery Conservation and Management Act (MFCMA) of 1976, each of eight regional fishery management councils has primary responsibility for preparing a fishery management plan (FMP) for fisheries in its region in federal waters. In developing a FMP, a council will hold public scoping meetings and receive the views of interested parties. Since the basic objectives of an FMP are set by a council very early in the process, it is important to become involved early. The council may hold additional meetings to solicit and review additional information. It then prepares a draft FMP and sends it forward to the Secretary of Commerce through the Regional Director of the National Marine Fisheries Service. The public then has another opportunity to comment on the proposed FMP in writing or at pubic hearings. If the Secretary of Commerce finds that the FMP complies with the Act, regulations are issued. If the Secretary of Commerce determines otherwise, then the FMP is returned to the council for revision. The Act requires councils to review their plans regularly. Amendment of an FMP itself follows the same process as initial approval of an FMP. To learn about current FMPs, proposed FMPs, or the FMP process, it is best to locate and contact your regional Fisheries Management Operations Division and the Marine Fish Conservation Network. (See Appendix 3.)

ENDANGERED WILDLIFE

In the late 1960s, as concern for the environment grew, Congress passed one version of what became the Endangered Species Act of 1973. The Department of the Interior's Fish and Wildlife Service has primary responsibility for listing and other aspects of the federal endangered species program. However, prior consultation with and approval by the Secretary of Commerce is required for listing marine species. Citizens can petition the Secretary of the Interior to add or remove species from the list of those protected by the Act. If the petition presents adequate information, the Department of the Interior requests public comment. The ESA also has specific provisions for citizens to file lawsuits for compelling federal and state agencies to comply with the statute. For more information on federal endangered species programs, contact the Office of Protected Resources in the National Marine Fisheries Service and the Office of Endangered Species in the U.S. Fish and Wildlife Service. (See Appendix 2)

MARINE MAMMAL PROTECTION

In 1972 Congress passed the Marine Mammal Protection Act. This law prohibits almost every action that can harm marine animals (any form of harming is referred to as a "taking" in the law), but also includes several exceptions. Importantly, most of these exceptions require anyone who wishes to "take" a marine mammal to first get a permit from the National Marine Fisheries Service or the Fish and Wildlife Service. These federal agencies publish notices in the *Federal Register* of each permit application they receive and invite public comment on the proposed taking. This is an excellent opportunity for involvement. In cases likely to be controversial, they sometimes hold public hearings at which you can present your views.

MARINE DEBRIS

Under the MARPOL Treaty, U.S. law now prohibits vessels from disposing of plastics and regulates the distance from shore that other types of garbage may be dumped in U.S. waters. (See Appendix 1.) The U.S. Coast Guard is responsible for enforcing MARPOL Annex V in U.S. waters and citizens can help the Coast Guard by reporting any dumping violations they witness. Contact the NOAA Marine Debris Information Office (Appendix 4) for copies of the reporting form and call the Coast Guard office nearest you to report a violation. Check your phone book or contact the Washington, DC headquarters. Citizens can also participate in beach cleanups, collecting trash, and recording the information for data cards. Contact the EPA in your region for more information. (See Appendix 2)

NOTES

LEGISLATIVE ACTION

"People all over the world are demanding more from their leaders–and getting it.
In America, poll after poll tells us that the people are way ahead of the politicians.
The most important step we can take is to tell our leaders that if they don't protect
the environment, they won't get elected." — **Robert Redford**

As concerned citizens, divers have the power to ensure that our elected officials put marine conservation at the top of their legislative agendas. We can make the democratic process work for us at the federal level as well as at the local and state levels, whether it involves influencing a member of Congress or informing the White House of our concerns. As you discover how the legislative system works, you become part of the process itself by raising questions and concerns and being heard. Understanding the way in which the federal government works is the first step. The following is an overview of the legislative process.

The United States Congress and state legislatures play significant roles in governmental conservation efforts. Most federal and state programs have as their basis a law that has been approved by the legislature and signed by the president or governor, as the case may be. A law that establishes a program, defines its purposes and procedures and authorizes funding is called authorizing legislation. Examples are the federal Endangered Species Act or a state law establishing a coastal zone management program.

The legislative process often begins with the introduction of a bill. Bills introduced in the House of Representatives are given a number preceded by "H.R.," for example, H.R. 2647. The numbers for Senate bills are preceded by "S.," for example, S. 1189. Generally, committees and sub-committees hold hearings and amend an introduced bill before sending it to be voted on by the full House or Senate. Both must agree upon a common text in order for a bill to be sent to the president for his signature. Authorizing legislation is not static; generally, it is in force for a limited number of years.

Congress reauthorizes legislation periodically, and often amends the authorizing legislation in the process.

In order to implement a program created by authorized legislation, federal and state agencies need funding. The authorizing legislation only sets upper limits on the amount of funding that may be made available to an agency. Separate legislation appropriating funding for a program often must be passed by the legislature. The amount of money appropriated by a legislative body can make the difference between an effective program and a program that exists only on paper.

Legislative bodies take other actions that influence the conservation of marine environments. For instance, a committee may hold investigative hearings on an emerging issue or on the administration of a particular program. Such hearings often lead to new legislation. Legislatures may pass resolutions that express the legislature's views regarding an issue to the administrative branch of government. These resolutions are not binding, but agencies generally have to keep them in mind for fear of provoking stronger legislative measures.

Finally, individual legislators may send letters to or meet with agencies if they are concerned that the law of the land is not being observed or is harming an interest of theirs. The legislative process is often long and tortuous. Nonetheless, individual citizens and organizations generally have several opportunities for making their views known. The earlier you become involved, the greater your effectiveness.

If you are interested in discussions on the floor of the U.S. House or Senate, you may wish to consult copies of the *Congressional Record,* available at any large library. The *Congressional Record* includes all speeches, statements, bills, and resolutions that were part of a particular day's business in the House and Senate. House and Senate business may also be viewed on the cable television station called C-SPAN. Your state's legislative business may be viewed on local cable television.

How to Communicate Your Views to Legislators

Although legislators will be most responsive to communications from their own constituents, they will often listen to well-articulated views of other citizens. There are several ways to convey your viewpoint to a member of Congress. One of the most powerful vehicles is letter writing. A thoughtful personal letter can change the mind of a legislator who would otherwise be unconcerned with an issue. There are a number of ways to make your letters most effective:

1. **The most effective letters are original and personal letters.** Use your own letterhead or stationery. A personal letter is far better than a form letter or a signature on a petition.

2. **Make your letter concise, no longer than two pages.** Write about only one subject in your letter. Whether you type or write your letter by hand, make sure your letter is clear and easy to read.

3. **Your letter should be timely.** Inform your legislators while there is still time for them to take action.

4. **Your first sentence should state where you live,** especially if you are a constituent.

5. If you are writing about a bill under consideration by the legislature, **refer to the bill by its name, or preferably by its number.** If you are writing about an action that an agency has recently taken, identify the date on which the action was taken. Senate Bills begin with "S." and House bills begin with "H.R." Therefore, your letter could begin with: "I am writing to urge you to co-sponsor S.___ or H.R.____."

6. **Describe your concerns or your position on the issue, and why you hold that position.** You do not have to present all the reasons for your position, only those which you feel are most important and most convincing. Try to show an awareness of how the proposed legislation would affect not just the environment, but also your community and other people's health and jobs.

7. **Be specific.** Tell the legislator what you wish done. Ask whether your legislator will support or oppose the legislation you are interested in. If you are requesting your legislator to investigate an agency action, explicitly ask that this be done.

8. **It is not necessary to mention the conservation organizations with which you are affiliated.** Members of Congress usually know where such groups stand on environmental

issues. They want to know your personal view on the issue. Use your own words and be sincere. Specify if you have expertise on the subject; your legislator will very likely appreciate professional perspectives.

9. **Write your own letter.** Do not send photocopies or preprinted postcards unless you absolutely cannot do otherwise.

10. **If your senators or representatives serve on any of the committees that have jurisdiction over the issues you are concerned about, send a copy of your letter to those committees as well.**

11. **Request a response.** For example, "I look forward to hearing how you vote on this bill," or "I look forward to hearing your views on this issue." Make sure that you include your address so they can respond.

12. **Thank the legislator for considering your views.**

13. **Do not threaten a legislator** with your vote or with your influence.

14. **If time is of the essence, send a Mailgram** through Western Union.

15. **Make sure your letter is correctly addressed.** Unless you know your member's room number, address your letters as follows:

The Honorable	The Honorable
(your Senator)	(your Representative)
United States Senate	U.S. House of
Washington, DC 20510	Representatives
	Washington, DC 20510

Addresses of state legislators are often found in your telephone book, or you can consult local libraries. An excellent resource is the *Legislative Directory*, published by the American Gas Association (1515 Wilson Blvd., Arlington, VA 22209). This compact handbook contains names, addresses, phone numbers, and photos of all congressmen and senators, a complete list of federal agencies, committees, subcommittees, and their staff members, plus other useful information.

In addition to writing your representatives in Congress, you can also share your views and concerns with the president and the vice president. Letters to either leader should include information similar to that found in letters to Capitol Hill. These letters should be addressed as follows:

President _____
The White House
1600 Pennsylvania Avenue
Washington, DC 20500

Vice President _____
Old Executive Office Building
Washington, DC 20501

16. **Follow-up letters are important.** If you do not receive a reply from your representative within a few weeks, send another letter. If you do receive a reply stating that your legislator agrees with your views and intends to support appropriate conservation legislation, write a thank you note. If you receive a non-committal reply that does not adequately address the points in your original letter, write another letter and politely readdress your concerns. In most cases, your representatives will respond more directly to second letters.

Telephone calls to a member of Congress are of limited value unless you can get a number of other people to make calls as well. If you do call, be specific and courteous. To find out the telephone number of your member, call the Capitol switchboard: (202) 224-3121. Telephone numbers for state legislators are often found in your local telephone directory.

Knowing *when* to write is as important as knowing *how* to write. If you are writing about a specific piece of legislation, the most opportune times for writing are when the bill is introduced, when a committee holds hearings on the bill, or when the full legislature will be voting on the bill. Many organizations listed in Appendices 2 and 3 can keep you informed on issues of interest.

If your legislator has acted as you have requested, express your appreciation. If nothing else, such a letter will show that you are paying attention. The legislative process is full of surprises: a scheduled vote on a bill may be delayed at the last second by the parliamentary moves of an opponent. Even legislative staff who may be quite familiar with the process can be caught by surprise.

One of the greatest difficulties confronting a citizen who does not live in Washington, DC or in a state capital is obtaining timely information. But there are several ways to overcome this difficulty. One of the best is to inform an organization that is working on your issue that you wish to be actively involved in their efforts. Also, legislators often maintain offices in their home districts or states. If you have a question about a piece of legislation, give these offices a call. Some states have toll-free hotlines to keep citizens informed of pending state legislation.

If you do not know who your legislative representatives are, contact your local public library, or purchase the *Legislative Directory* available from the American Gas Association. Also, agencies responsible for implementing specific legislation often know about pending legislative action and which legislators are most involved. Public affairs divisions in state and federal agencies are a good place to start in obtaining such information.

LOBBYING

Another excellent way to communicate your concerns to legislators is by lobbying. Nothing impresses members of Congress as much as citizens willing to make a personal visit, and legislators are often influenced more by face-to-face interviews with concerned citizens than by paid lobbyists.

You do not have to travel to Washington to meet with your representatives, as they return home to their local or state office on weekends and major holidays. They also appear at town meetings to gather constituent views and support. These are wonderful opportunities for you as an individual, or as a representative of a group, to expression your opinions.

To be successful at lobbying, consider the following:

1. **Do a background check on the legislators with whom you intend to meet.** Learn as much as you can beforehand about their position statements, voting records, committee assignments, and other pertinent information. These insights will help you select the legislators who may be the most influential in your issues and will help you prepare your approach. *Project Vote Smart*, published by the Center for National Independence in Politics (129 NW Fourth St., Suite #204, Corvallis, OR 87330, phone (503) 754-2776, FAX (503) 754-2747) is full of helpful information about U.S. congressmen and senators including their biographical backgrounds, campaign finances, performance evaluation, special interest

groups, position statements, local election office addresses and phone numbers, and committee assignments. You can also obtain information about legislators by calling the Voters Research Hotline: 1-800-622 SMART.

2. **Make an appointment.** Call the district or state office and request a meeting during the next recess break when your legislator is at home. Many members of Congress are in their districts Friday, Saturday, Sunday, and Monday. The appointment secretary will probably want to know the nature of your business. Limit your agenda as much as possible, preferably to one topic. You can telephone your senators or representatives through the Washington, DC, U.S. Capitol switchboard: (202) 224-3121. Many legislative offices will accept "walk-in" visits, but you will most likely have to speak with an aide, not the legislator.

3. **Come prepared.** It is always better to come alone or in small groups. If you come alone, identify which interests you represent. If you come with others, plan ahead who will be the spokespersons to represent different groups. Bring materials to leave with your legislator which recap your concerns and your views.

4. **Keep it brief and to the point.** Since a legislator's schedule is busy, keep your visit short. Decide beforehand who will say what and bring notes with highlights of the key issues to help keep the discussion on track. If your legislator strays off the subject and seems to waste time, politely but firmly get him or her back to the issue at hand. Do not overstay your allotted time. It may help to hold a "rehearsal" meeting beforehand.

5. **Be specific.** Name the bill in which you are interested and/or the action which you want your legislator to take. Try and find a local angle on national or regional issues.

6. **Look and act professional**. Be on time, dress formally (conservatively), and be polite. Introduce yourselves cordially, and state your reasons for requesting the meeting. Never threaten your legislator or pretend to wield a lot of influence if you do not.

7. **Follow up.** Follow up with a thank-you letter, and provide any information which you promised during the meeting. Use this opportunity to continue to build a relationship with your elected official and his or her staff. If you do not get the response you want, be polite but persistent, and do not be vindictive. There may be other issues which your legislator will act upon favorably in the future.

Although lobbying is a distinctly personal activity, the Audubon Society in Washington, DC runs workshops to teach people how to lobby on a various issues. During a typical three-day workshop, participants learn the ins and outs of lobbying from experts in the field, and meet with congressmen and senators on Capitol Hill. (For more information, call the National Audubon Society, listed in Appendix 3.) In addition to the National Audubon Society, there are a number of other groups which specialize in networking, lobbying, and legislative action:

The Advocacy Institute
1730 Rhode Island Avenue NW
Washington, DC 20036

Inform
381 Park Avenue South
New York, NY 10016

Concern
1794 Columbia Road NW
Washington, DC 20009

The Learning Alliance
494 Broadway
New York, NY 10012

Renew America
1001 Connecticut Ave. NW
Washington, DC 20036

INTERACTING WITH THE EXECUTIVE BRANCH OF GOVERNMENT

While legislatures are responsible for providing the legal framework in which marine conservation programs operate, executive agencies are responsible for implementing these programs. (See Appendix 2.) Environmental laws generally place the authority and responsibility for implementing a program in the chief executive officer of a department or agency, who then delegates the authority to the chief of the division most expert in the area. For instance, the Magnuson Fishery Conservation and Management Act vests great authority in the Secretary of Commerce, but this authority has effectively been delegated to the Assistant Administrator for Fisheries in the National Oceanic and Atmospheric Administration (NOAA). Indeed, many decisions under this Act have now been delegated to the Regional Directors of the National Marine Fisheries Service.

Depending upon the gravity and controversy surrounding a decision, these individuals will have to clear decisions informally with the higher levels of government. Thus, besides communicating your views to the division chief of the agency with responsibility for an issue, you probably need to contact these other levels of government if an issue has generated a lot of debate.

The administration of some marine conservation programs can be even more complicated. For example, under the Magnuson Fishery Conservation Management Act, the eight Regional Fishery Management Councils are composed of representatives of state government, the fishing industry, and other interest groups. Each council is responsible for developing fishery management plans for major fisheries in the federal waters of its region. Although the Secretary of Commerce must ultimately approve of a plan, the councils exercise considerable influence over the management of fisheries. Greater involvement by the general public is needed at these hearings. Many decisions that directly affect the health of our coastal waters and habitats are made by local planning commissions and town councils. Since the operation of these local government agencies varies so much, general descriptions are of little value. Nonetheless, the general rules discussed above apply at the local level as well. Above all, the better informed and thoughtful your participation is, the more effective you will be in influencing local decisions.

Obtaining Information about Agency Actions

Generally, local, state, and federal agencies must base their decisions on a record composed of information gathered by staff and of comments from the public and other agencies. Federal law and some state laws require that agencies conduct reviews of the environmental impact of major actions. The National Environmental Policy Act requires that a federal agency prepare an Environmental Impact Statement (EIS) if the agency decision or action will have a significant effect on the human environment. An EIS is an analysis of the risks and benefits of a proposed activity and alternatives to the proposed activity. In the past, federal agencies have had to prepare EISs for such activities as offshore oil and gas lease sales, port and harbor construction, fishery management plans, and state coastal zone management plans.

This environmental review process begins with public scoping hearings called by the federal agency. At these hearings, the public can identify concerns arising from the proposed action; indeed, unless a concern is identified at this stage, it may not be addressed in an EIS. The agency later releases a draft EIS, which is distributed to other agencies, interested individuals, and organizations. The agency conducts public hearings, generally in the geographical areas most affected by the proposed action, and accepts oral and written comments on the draft EIS. After the comment period ends, the agency reviews the comments and prepares a final EIS. This forms the basis for the final agency decision.

Notices of draft environmental impact statements, public hearings, and public comment periods are always published in the *Federal Register.* (See Chapter 19) However, the best way to insure that you are informed about such proposed agency actions is to contact the agency responsible for issues of interest to you or organizations that regularly participate in such decisions.

Many decisions made by local, state, and federal agencies do not require such extensive review and are therefore more difficult to track. Decisions regarding permits, as for dredging wetlands areas, are reviewed by federal and state agencies and may be subject to a public review period. Once again, the best way to be made aware of these decisions is to get your name placed on the appropriate mailing list. Notices about local decisions are often posted in public libraries or in local newspapers.

State agency actions are guided by the rules, laws, and policies approved by the state legislature or governor's office. Individuals and organizations can take an active role in helping create these rules. Often, public meetings and hearings are held, usually in several locations around the state, to facilitate public input. As with the other actions discussed above, the best way to stay involved is to be on that state agency or department mailing list. If you believe that a proposed decision by an agency is of great enough concern, you may wish to contact your legislator and ask for assistance. In doing so, be specific by identifying the agency responsible for the action, the action itself, and the date on which the action occurred. Once again do your homework, think through your arguments, and be polite, but persistent.

MAKING CONSERVATION
A WAY OF LIFE

"The greatest natural resource on the planet is the human race itself. And, in a world in need of ever-increasing care and protection, the full potential of every individual has never been more in demand." — *Dr. Norman Meyers*

More and more, people are beginning to realize that everything we do affects the environment. The NIMBY ("Not In My Back Yard") principle, which had historically been used to justify environmentally irresponsible and selfish actions, is obsolete. We can no longer pollute "other people's back yards" with impunity, because pollution today has a boomerang effect. If we dump harmful substances down the drain, down a sewer, into a gutter, on a lawn, in the air, or on the ground, they eventually end up in and harm our collective back yard...the marine environment.

Achieving success in marine conservation–that is, preventing marine and atmospheric pollution, reducing waste, preserving valuable habitats and ecosystems–therefore necessitates changes in both personal lifestyles and in industry. These social and economic changes mean that we must be willing to sacrifice some modern conveniences for alternatives which are less stressful to nature. All too often technology is depicted as the arch-enemy of nature, but if it used responsibly, technology can further conservation. There is enormous potential for success when technology, ecology, and society all work in harmony.

The following are examples of ways in which all of us can use less energy and natural resources in our houses, workplaces, and public areas, producing less stress and waste in the environment. (Some of these tips are recommended by Clean Ocean Action.)

CONSERVATION AT HOME

1. **Practice the Three R's.** One of the most direct things people can do in their homes to help the environment is to practice

the three R's–Reduce, Reuse, and Recycle. Buy only products that are necessary and durable. Try to reuse items as much as possible, or give them to others who can reuse them. Discard products only as a last resort. Make sure that you buy products which are recyclable and have packaging which can also be recycled.

2. **Save energy.** A large percentage of the electricity we use around our homes goes into lighting. Using lighting intelligently and taking advantage of efficient lighting technologies can save fifty to ninety percent of that energy. This will lower your electric bill, reduce acid rain and nuclear wastes, and cause fewer greenhouse gases to be emitted. Whenever possible, use natural daylight or fluorescent bulbs instead of standard incandescent bulbs (fluorescent bulbs produce the same light as regular bulbs, but use only one quarter of the energy). Use outdoor lights only when they are needed.

3. **Save heat.** The average American household expends fifty to seventy percent of its energy on heating and cooling. Keep thermostats at 68° or lower. Minimize or eliminate the use of air conditioners (air conditioning reduction saves energy and also lowers CFC emissions). Keep lights and appliances off when they are not in use. When heating with wood, use wood that gives the most heat, such as black birch, hickory, live oak, locust, northern red oak, rock elm, sugar maple, and white oak.

4. **Save water.** The water that drains away from showers, sinks, and toilets all ends up in the same place–the sewage treatment plant or septic system. Processing this waste water uses up energy and money. A family of four people can easily use 1,500 gallons of water per day maintaining an average household and a garden. To conserve water check outside and inside to ensure that your plumbing fixtures do not have leaks. A leak can waste hundreds of gallons of water a year. Other ways of conserving water include taking quick showers, not using the toilet as a waste basket, turning off the tap when brushing your teeth or shaving, and running the washing machine and dishwasher only when full. Use water sparingly outdoors as well. Think conservation when it comes to washing a car, hosing down a sidewalk, watering a lawn, or filling a swimming pool.

5. **Use a microwave oven.** Microwave ovens use about the same amount of energy as conventional ovens, but they cook most foods in less than half the time, which saves half the energy.

6. **Eliminate toxics.** The EPA has classified four types of dangerous materials found in households: those which are flammable (can catch fire easily), corrosive (can dissolve metals or burn skin), reactive (dissolve violently with water or other substances), and toxic (poisonous). Find alternatives to dangerous products, such as drain cleaners, disinfectants, and cleaners containing phosphates. Many of these products are harmful to your health as well as to the environment. Most people have products in their home which they do not know are potentially dangerous and which require special disposal. For example, the nickel, cadmium, mercury, and alkaline in batteries are highly dangerous to humans, animals, and fish. If you put a battery in a wood stove or open fire, it will explode. Used mercury batteries should be taken to a hazardous waste center!

7. **Check to see that your drinking water is not contaminated.** In towns with a population greater than 10,000, any system supplying more than fifteen homes is required by law to be tested for bacteria, organic chemicals, pesticides, and other contaminants at least once a year. If your water comes from a public supplier, your water bill will indicate the local water company, municipal board, or regional authority. You have a right to ask the supplier for copies of their water test results. If the pollution level exceeds the EPA standards, ask your local or state health department to conduct a test to see whether the contamination stems from your own home or from the supplier. If it stems from the supplier, notify the EPA or other appropriate authority.

SHOPPING

1. **Patronize eco-friendly manufacturers and businesses.** Many businesses are striving for a good environmental image. They are committed to supporting and promoting social and environmental changes. Some use biodegradable and recycled products and containers. For example, Chrysler Corporation, Ford Motor Company, and GM have formed a

partnership to conduct research on recycling and disposing of materials from used automobiles. Others use only natural ingredients in their products, and do not use animals for testing (such as the Body Shop and Kiss Colors).

Some companies donate a portion of their profits to environmental projects and organizations. The Nature Company, which produces a host of ecologically sound products, donates money to the Nature Conservancy. Esprit, the clothing manufacturer, uses no harmful chemicals or synthetic fibers, and also supports economically depressed areas and vanishing cultures. For example, Esprit's reconstituted glass buttons are made from old bottles by handicraft workers in Ghana. MacDonalds Corporation has been recognized by environmentalists for developing a list of sixty-two solid waste reduction initiatives independently and in conjunction with the Environmental Defense Fund. The Royal Bank of Canada donated almost a million dollars to fund the construction of the building that houses the Center for Biological Diversity in Georgetown, Guyana.

2. **Choose environment-friendly products.** "Environment-friendly" products are becoming increasingly popular. Choose products with natural ingredients. Avoid cosmetics, cleansers, and other household products which contain caustic substances, such as formaldehyde. Avoid products which are based on materials taken from threatened environments or endangered species. Avoid highly processed products which necessitate environmentally destructive processes or involve wasting energy. When possible, choose plain or natural substitutes for products which contain dyes or chemical fragrances (such as some toilet paper).

3. **Choose products which can be recycled or reused.** Consumers can buy everything from recycled steel cleaning pads to non-radioactive smoke detectors, to pencils handmade from gathered twigs. However, it is important to watch out for misinformation about products which claim to be degradable and recyclable. Make sure the products you buy are not more harmful to the environment because they promote "one use" mentality or are not readily recycled.

4. **Weigh the pros and cons of a product.** For example, using cloth diapers may seem less stressful to the environment

than disposable plastic diapers, but if you factor in the detergent, bleach, softening agents, and energy needed to clean cloth diapers (plus the fuel that is burned by diaper service delivery vehicles), many conservationists consider disposable diapers the better choice.

5. **Avoid products which have been treated with harmful pesticides or harvested irresponsibly.** All tuna should be labeled as dolphin safe. Many people are eliminating meat from their diets because it contains high saturated fat and additives, but also because the beef industry uses enormous natural resources. (Statistics to support this statement include: it takes twenty-five gallons of water to produce a pound of wheat, but twenty-five *hundred* gallons of water to produce a pound of meat; eighty-five percent of U.S. topsoil is lost directly from raising livestock, and 260 million acres of prime U.S. forest had to be cleared for cropland to feed cattle; 1,000 species a year are becoming extinct due to destruction of tropical rain forests for uses relating to livestock.)

6. **Buy only what you need.**

7. **When purchasing clothing, buy articles that can be interchanged to make new outfits.** You can also recycle clothes by giving them to the Salvation Army, passing them on to other people (in the U.S. and in other countries), and by using thrift shops.

8. **Voice your opinion to retailers and manufacturers.** You can put your money where your conservation ethic is. Make your anger known if a company is operating or making products which are ecologically irresponsible. You can request a phone conversation with a company executive, write a letter, or combine your effort with other concerned people and create a petition. If you do not get an acceptable response, notify local environmental and business organizations, and if necessary the media. Above all, use your most powerful weapon—your right to boycott. When retailers and manufacturers see that their actions are hurting profits, they are most likely to change. One such success story took place in 1989, when the American apple industry stopped using Alar, a chemical sprayed on apples to enhance color and prolong shelf life. The action was not the result of government

intervention, but of environmental group pressure and, more important, consumer power in the marketplace. People refused to buy a product that jeopardized the health of humans and the environment. To help shoppers identify environment-friendly products, the Council on Economic Priorities (CEP) offers a useful handbook entitled *Shopping for a Better World—A Quick and Easy Guide to Socially Responsible Supermarket Shopping* (1989). CEP also created a Seal of Approval which is awarded to environment-friendly companies and products, such as Johnson Wax, the first company to ban CFCs.

TRANSPORTATION

1. **Drive less.** Cars burn fossil fuels and spew waste gases into the atmosphere, which contributes to acid rain. According to *1,001 Ways to Save the Planet*, "The car is the most environmentally-polluting form of transport that's ever been invented." Instead of using your car, car-pool, use mass transit, cycle, walk, or skate. Reducing the use of cars not only conserves fossil fuels, it also lessens the need to drill and ship oil.

2. **Buy a fuel efficient car.** If using a car is mandatory, use one with a mileage rating of 30 miles per gallon or more. Using more fuel-efficient cars reduces oil spills and the need for off-shore drilling.

3. **Idle less**. Turn off your car if it will be idling more than one minute. Clean Ocean Action research revealed that in New Jersey alone, idling cars stuck behind open draw bridges waste enough fuel each day to drive a 30 mile per gallon car around the world 26 times!

4. **Check your tires.** Poorly-inflated tires can result in up to 10% fuel loss. Radial tires are more fuel-efficient than cross-ply. Ask your dealer or county solid waste coordinator about recycling old tires.

5. **Check for leaks.** Leaky gaskets, radiators, and crankcases drip staggering amounts of oil, anti-freeze, and other automotive toxins onto roads, parking lots, and pavements, all of which eventually lead into waterways. Millions of gallons of oil leach into U.S. waterways from non-point sources every year.

6. **Keep your car in tune.** Keeping your car properly tuned up results in 5% better fuel efficiency and cleaner emissions. Frequent automotive checks will also detect leaks and other pollution problems.

7. **Change oil.** Frequent oil changes with high-quality oil according to manufacturer's instructions will give better mileage and a cleaner burning engine.

8. **Recycle used oil, antifreeze, and automotive fluids (brake and power-steering fluids, engine-radiator flushes).** Both oil and anti-freeze should be recycled–never poured down storm drains or dumped in the ground. By law, service stations that do oil changes must also accept used motor oil. Antifreeze can be recycled at household hazardous waste disposal days (call your county solid waste coordinator for details). You can purchase non-toxic anti-freeze at most automotive stores.

9. **Avoid using the air conditioner.** Car air conditioners burn tremendous amounts of gas and release toxic chlorofluorocarbons into the atmosphere.

10. **Dispose of used tires correctly.** Do not leave tires lying around. They are breeding grounds for rodents and mosquitoes. They are also a serious fire hazard, and if burned, a source of heavy air pollution. Contact your local collection center for special recycling instructions.

LAUNDRY AND CLEANING

1. **Read the labels.** Many household products contain ingredients which are hazardous to human health and the environment. Chemical ingredients to avoid include:
   trichloroethylene–used in polishes, aerosols, and waterproofing;
   nirtobenzene–found in polishes;xylene–found in cleaners;
   methylene chloride–found in cleaning and polishing solutions and paint strippers;
   formaldehyde–found in cleaning solutions, disinfectants, and building materials;
   chlorine–found in bleach and scouring powders; and
   ammonia–found in glass cleaners and all-purpose cleaners.

2. **Avoid aerosols.** Many still contain ozone-depleting CFCs. Alternative propellants, such as butane and propane, may adversely affect the heart and central nervous system.

3. **Buy phosphate-free, biodegradable soaps and detergents.** Phosphates are directly related to algae blooms which degrade water quality. Use soap products such as Ivory Snow, and add Borax to the wash to help suspend the soap. As an alternative to chlorine bleach, try adding washing soda to one cup of lemon juice. Use Murphy's Oil Soap directly on colorful products as a stain remover.

4. **Avoid dry cleaning.** Dry cleaning uses toxic chlorinated solvents. If it is necessary to dry clean, air clothes outside for a few hours before bringing them indoors. If there is a noticeable odor from your clothes, change cleaners, as they may be using an excess of solvents.

5. **Buy water-efficient appliances**. When you are in the market for a new washing machine or dishwasher, shop for one that is advertised as water efficient. Operate appliances when they are full to reduce energy demands and save water.

6. **Avoid commercial metal polishes.** They contain phosphoric and sulfuric acids and ammonia. Silver flatware can be cleaned by boiling in baking soda and salt. Using a paste of baking soda and water, you can polish silver and stainless steel. For brass, use equal parts of salt and flour, with a little vinegar. A mixture of lemon juice or hot vinegar and salt will polish copper. Chrome can be polished with a little rubbing alcohol and white flour.

7. **Shampoo carpets carefully.** The residue from commercial solvents has been linked to respiratory infection. It also contributes to indoor air pollution. Steam cleaning is safer, and for freshening and light cleaning, try sprinkling corn starch and then vacuuming.

8. **Use alternative cleaning solutions.** Use Bon Ami instead of scouring cleansers containing chlorine. For tough stains, use half a lemon dipped in Borax, and rub. Murphy's Oil Soap and Borax are good for general cleaning. For a glass cleaner, first use alcohol to remove wax residues left from commer-

cial glass cleaners. Follow up with a mixture of 50% white vinegar and 50% warm water.

9. **Dispose safely.** Almost every household contains toxic materials. Never dump caustic chemicals on the ground or in toilet bowls and sewers. You must dispose of these items safely, during designated hazardous waste disposal days in your community. You can call your County Solid Waste Coordinator to find out when such events are conducted.

10. **Service and dispose of your refrigerator or home air conditioner responsibly.** Most home refrigerators use CFCs as the coolant, and most home air conditioners use HCFCs. If repairs are necessary, make sure that the service company captures and recycles the coolant. When disposing of these items, check with the manufacturer or local service shop about disposal methods which avoid releasing toxins.

11. **Use fiberglass or cellulose insulation.** CFCs are used to manufacture foam insulation, although manufacturers are rapidly switching to alternatives.

LAWN AND GARDEN

1. **Check the pH.** Do not add anything to your lawn or garden until you have your soil's pH tested for acidity and fertility. This can be done through your county Agricultural Extension Service or by purchasing a kit at your local garden center.

2. **Aerate your soil and remove dead organic material by raking.** This produces a stronger lawn and reduces the need to use fertilizers or pesticides.

3. **Use organic fertilizers to condition the soil.** Chemical fertilizers (which run off lawns into waterways) produce chemically-dependent lawns which require more and more chemicals to remain healthy. Some brand-name organic fertilizers are Espoma, Earthworks, Ringer Lawn King, Fertell, Earth-Rite, and Sustain. For a greener lawn, natural sources of nitrogen, such as composted manure, blood meal, and cottonseed meal, can be used.

4. **Do not overfertilize your lawn.** Overfertilized grass roots become lazy and remain near the surface, where they require more water and are more exposed to extremes in weather. Bonemeal and rock phosphate will aid in building strong roots.

5. **Plant drought-tolerant grass.** Cultivate and mulch your gardens to help retain moisture and keep weak growth down. Water only when the soil is dry, but water thoroughly to encourage deeper root growth, which reduces the need for excessive watering. Garden centers sell water meters which indicate when water is needed. To discourage disease, water in the morning so that lawns and plants will be dry by evening.

6. **Mow high and often.** Mowing high and often reduces stress on grass and helps retain moisture, shades out weeds, and keeps soil cool. Mow without a bag and leave your grass clippings, which will fertilize naturally and shade the soil.

7. **Start a compost pile.** Excessive grass clippings, leaves, weeds, kitchen waste, etc. can all be used to create a nutrient rich fertilizer to till into gardens.

8. **Avoid using chemical pesticides.** Pesticides run off into waterways, kill beneficial insects, and are long lasting. Large insects (e.g., Japanese beetles) can be hand picked from gardens and dispatched in soapy water or alcohol (do not use gasoline or other toxins). Biological non-toxic controls are available at garden centers.

9. **Avoid lawn services that use chemicals.** Run-off from these chemicals contribute to non-point source pollution. Furthermore, despite lawn companies' claims that treated areas are safe after twenty-four hours, dangerous pesticides remain long after application.

10. **Plant trees.** They contribute oxygen, remove carbon dioxide, reduce erosion, and provide shade.

CONSERVATION AT THE OFFICE

1. To make it easier to **recycle at your desk,** have separate trash cans for recyclable items vs. non recyclable trash.

2. When photocopying documents and memos, **use both sides of the paper.**

3. **Use routing slips** to circulate your internal memos and reports instead of making multiple copies of the same document.

4. **On wastepaper that has copy only on one side, use the other side for scratch paper.** Even envelopes can be used for notes.

5. **Reuse envelopes** when sending letters, documents, and materials out. Just put your label over the old address.

6. When shopping for stationery or paper for printing, whenever possible **choose paper that has at least ten percent post-consumer recycled content.**

7. Fill the paper feeder of your copier or laser printer with one-sided, already used paper.

IN INDUSTRY

The "greening" of corporate America has been moving forward slowly, but changes can definitely be seen in industry policies and products. Federal and state environmental standards are becoming more stringent, and the public is becoming more demanding. An example of this can be seen in the decline and demise of companies which sanction or practice environmental degradation. Avtex, formerly one of the three largest rayon producers in the U.S. and the sole supplier to NASA of carbonized rayon used in rocket nuzzles had to shut down because of unresolved pollution problems and its failure to meet state environmental standards. Prior to its shutdown, Avtex knowingly contaminated areas in and around its factory headquarters, spewing pollutants from smokestacks into the air, which endangered the health of humans and natural ecosystems.

Also, manufacturing rayon destroys wildlife and biodiversity. Rayon is derived from the pulp of trees, and the largest stands of old-growth trees are logged first. In 1992, Courtaulds Fibers, now one of the two largest rayon producers, introduced Tencel–a silky fabric similar to rayon. Wood pulp from eucalyptus trees, harvested every seven to ten years, provide the fiber for Tencel. The trees are specifically cultivated for this purpose. No wilderness areas or old-growth forests are logged for raw materials. The wood pulp or cellulose, is converted to fiber by using solvents which are recycled. Any solvents dispersed in the wastewater during manufacturing are benign. Tencel can be hand washed, eliminating the need for dry cleaning.

It is success stories like this which pave the way in conservation. People who own or have decision making power in industries and businesses can also promote conservation by:

1. choosing raw materials and methods of manufacturing that are least stressful to the environment;

2. insisting that every product reflect or at least approximate its actual cost, not only the direct cost of production and shipping, but also the costs to air, water, and soil, the cost to future generations, the cost to worker health, and the cost of waste, pollution, and toxicity. Synthetic "throw away" products may seem cheap, but environmentally they may carry an astronomical price!

3. manufacturing or using products which can be recycled or reused;

4. not using poisonous chemicals in manufacturing, cleaning, or other activities;

5. using procedures that minimize production of waste materials;

6. donating a portion of corporate profits to environmental funds and projects;

7. developing a good rapport and open dialogue with employees, community, and citizen groups to discuss ways of protecting and enhancing the local environment;

8. using hang tags, brochures, and advertisements which promote conservation, such a "Made in the USA–protect the earth"; and

9. providing education to workers to reduce risks to themselves and their environment through safe handling and work practices.

CONSERVATION IN THE 21ST CENTURY

"The future is discretionary. We have the ability to alter it."

— *Vice President Al Gore*

The marine environment has been the epicenter of life on this planet for over three billion years. However, explosive growth in human population and industry in this last century has encroached upon and altered the natural world at a rate which is unparalleled in history. As we approach the twenty-first century, scientists predict that humankind and the environment are on a disastrous collision course in which the earth will eventually no longer be able to replenish its resources as quickly as we are depleting them. Fortunately, we have the ability to avert that disaster.

One of the greatest challenges facing modern man is controlling and treating waste materials. Throughout history, wastes were disposed of by being dumped into oceans, lakes, rivers, and streams, buried in the ground, or by the burning. With the passage of the Resource Conservation and Recovery Act in 1976, and the Superfund in 1980, however, the management and disposal of solid and hazardous wastes have been subject to tight federal and state regulation (see appendix I for federal legislation). To meet the federal and state requirements, the scientific and engineering communities began developing new technologies and variations on the existing waste disposal methods.

These methods include:

1. thermal treatment: rotary kiln incineration, liquid injection incineration, fluidized bed incineration, infrared incineration, and plasma torching;

*The marine environment
contains the most
diverse and valuable
ecosystems on earth.*

–Photo: Stephen Frink

2. biological treatment in which wastes are introduced to a bio-mass of microorganisms (and often oxygen and nutrients as well) which metobolize the soluble organic components;

3. chemical and physical treatment strategies: dechlorination, UV oxidation, separation of waste components, soil washing, activated carbon absorption, soil flushing, vucuum extraction, radio frequency heating, and solidification/stabilization.

Our approach to conservation must take a dramatically new course in the twenty-first century. In the 1970s, for example, environmentalists identified the need for pollution management. They advocated strict controls on the release of dangerous toxins. But the controls which were established proved to be insufficient, largely disregarded by industry, and poorly enforced. Twenty years later,

Photo: Mort and Alese Pechter –

"The decisions we make or do not make, the actions we take or do not take will have repercussions for generations to come."

Life has existed in the sea for over a billion years.

–Photo: Keith Ibsen

toxic materials are still in abundance in the marine environment. It has become apparent that a *damage management* approach to the environment cannot succeed, and that conservation efforts must focus on *damage prevention.*

Such an approach necessitates changes in human attitudes and lifestyles. Generations of Americans have been nurtured on the belief that technology and wealth are the omnipotent forces of life, that the fabric of our planet is there to be exploited by whomever is prepared to pay the highest price. If conservation is to succeed we must downscale our material comforts and use technology wisely.

On the global scale, this means that people in developed countries will have to be willing to sacrifice some of the conveniences and creature comforts which modern technology has made possible. It also means that developed nations must assist undeveloped nations with their social, economic, and environmental management problems.

On the smaller scale, it means that each of us, individually and collectively, must enact changes to protect our water planet. There are numerous avenues to help us learn about marine ecology and promoting conservation–in academic, community, and political forums, in the water, and in everyday life.

It is natural for the conservation ethic to be spearheaded by the recreational users of the marine environment because these individuals have unique access to and reliance on the aquatic environment. Those who instruct and lead skin and scuba divers, boaters, fishers, and swimmers should promote conservation in all of their programs and activities. The commercial sectors of the marine recreational industries–manufacturers, retailers, training agencies, publishers, etc.–should make financial contributions to environmental projects and implement environmentally responsible practices in their workplaces.

Most of all, each of us must make a personal commitment to conservation. By being an environmental role model, we can inspire countless others to appreciate and protect the legacy of the sea. Conservation cannot be restricted to a single ecosystem, whether a rain forest or a coral reef. Preserving the beauty and health of our environment is inexorably linked not only to the essence of commercial and recreational activities, but ultimately, to human survival. Conservation is *now.* The decisions we make or do not make *now,* the actions we take or do not take *now,* will have indelible repercussions for generations to come.

REFERENCES

Bohnsack, J. "The Ecological Basis for Using Marine Fishery Reserves for Reef Resource Management." In *Proceedings of the Coral Reef Coalition Conference.* Key West, FL, 1992. Telephone conversation with author. 4 February 1993.

Broad, W. "Into the Abyss: New Robots Probe the Deep." New York *Times,* 9 March 1993. "Japan Plans to Conquer Sea's Depths." *New York Times,* 18 October 1994, C1.

Bulloch, D. *The Underwater Naturalist.* New York: Lyons & Burford, 1991.

Clark, R. Water *The International Crisis.* Cambridge, MA: MIT Press, 1993. "Endangered Marine Finfish–A Useful Concept?" Marine Fisheries And Endangered Species Committee Report. *Fisheries* (July/Aug. 1991). *Fish for the Future: A Citizen's Guide to Federal Marine Fisheries Management.* Washington, DC: Center for Marine Conservation, 1993.

Friedrich, H. *Marine Biology.* University of Washington Press, 1973.

Harms, V. *National Audubon Society Almanac of the Environment: The Ecology of Everyday Life.* New York: G.P. Putnam's Sons, 1994.

High, W. "Ghost Fishing Gear." Alaska Fisheries Science Center (Nov. 1990).

Johnson, F., and Stickney R. *Fisheries.* Kendall/Hunt Publishers, 1989.

Keen, E. "Ownership and Productivity of Marine Fishery Resources." *Fisheries* (July/Aug. 1991): 18-22.

Lawren, B. "Net Loss." National Wildlife (Oct./Nov. 1992): 47-52.

McDowell, J. "How Marine Animals Respond to Toxic Chemicals in Coastal Ecosystems." *Oceanus* 36 (2): 56-61.

Miller, G. *Living in the Environment.* 6th ed., Belmont, CA: Wadsworth Publishing, 1992.

Mitchell, J. "Our Disappearing Wetlands." National Geographic 182 (4): 3-45.

Naar, J. *Design for a Livable Planet.* New York: Harper & Row, 1990.

Norse, E.   *Global Marine Biodiversity.* Washington, DC: Island Press, 1993.

Odum, E.   *Ecology and our Endangered Life-Support Systems.* Sunderland, MA: Sinauer Associates, 1993.

O'Hara, K.   *A Citizen's Guide to Plastics in the Ocean: More than a Litter Problem.* Washington, DC: Center for Marine Conservation, 1988. "Our Living Oceans: First Annual Report on the Status of U.S. Living Marine Resources." Washington, DC: NOAA, 1991.

Pearce, J.   "Collective Effects of Development on Fish Habitats," *Oceanologica Acta* Vol. Sp, No. 11, 287-298. Personal communications with author. January 1993-September 1994.

Pearce, J.,   "A Review of Monitoring Strategies and
and   Assessments of Estuarine Pollution." *Aquatic*
Despres-Patanjo,   *Toxicology* (Nov. 1988): 323-343."No-Fishing Zones
L. Robertson, P.   Proposed Along the Atlantic Coast." *Tide* (Sept. 1992): 43-45.

Satchell, M.   "The Rape of the Oceans." U.S. News & World Report, 22 June 1992, 64-75.

Sherman, K.   "Where Have All the Fish Gone?" Nor'easter 4 (2): 14-19.

Simons, M.   "Mining Ravaging the Indian Ocean's Coral Reefs." New *York Times,* 8 August 1993. "Status of Fishery Resources off the Southeastern United States for 1991." NOAA Technical Memorandum NMFS-SEFSC-306, U.S. Dept. of Commerce. "Status of Fishery Resources off the Northeastern United States for 1992," NOAA Technical Memorandum NMFS-F/NEC-95, U.S. Dept. of Commerce.

Stead, E.,   *Management for a Small Planet: Strategic Decision*
and Stead, J.   *Making and the Environment.* Newbury Park, CA: Sage Publications, 1992.

Thorne-Miller,   *The Living Ocean–Understanding and Protecting Marine* B., and *Biodiversity.* Washington, DC: Island Press, 1991.

Catena, J.   "Biodiversity and Conservation of the Marine
Upton, H.   Environment." *Fisheries* (May/June 1992): 20-25.

Ward, F.   "Coral Reefs are Imperiled." National Geographic (July 1990): 115-132.

Weber, M.,   *A Nation of Oceans.* Washington, DC: Center for
Tinney, R.   and   Environmental Education, 1986.

Wells, S., and Hanna, N. *The Greenpeace Book of Coral Reefs.* New York: Sterling Publishing, 1992.

Wieland, R. Center *Why People Catch Too Many Fish.* Washington, DC: for Marine Conservation, 1992.

GLOSSARY OF TERMS

Abiotic Non-living.

Acclimation The adjustment to slowly changing new conditions.

Acid rain A poisonous precipitation created when exhaust gases from cars or factories mix with water in the air.

Adaptation Gradually evolved physical or behavioral changes in an organism that help it to survive.

Aerobic organism An organism that needs oxygen to stay alive.

Air pollution One or more chemicals in high enough concentrations in the air to harm humans, other animals, plants, or materials. Excess heat or noise can also be considered forms of air pollution.

Algae One-celled or many-celled plants that usually carry out photosynthesis in streams, lakes, ponds, oceans, and other surface waters.

Algae bloom A sharp increase in density of phytoplankton or benthic algae in a given area.

Alien species Species that migrate or are deliberately introduced into an ecosystem by humans. Alien species can take over and eliminate many native species.

Anoxic That which is devoid of oxygen.

Anthropogenic Arising from human activities, as opposed to natural origin.

Aphotic zone The lowermost zone of the ocean, which is without light.

Aquaculture Farming of marine organisms in ponds, lakes, tanks, or coastal areas.

Aquatic Pertaining to environments that contain liquid water.

Aquifer A porous underground stratum, such as limestone, sand, or gravel, bounded by rock or clay, which is shaped like an elongated pipe and stores groundwater.

Assemblage A group of creatures, which may or may not be of different species, which are all gathered together in one place, but are not necessarily interacting each other.

Assimilate To convert (food) into a substance suitable for absorption into the system.

Atmosphere The gases which make up the air surrounding the earth.

Benthic Living within, upon, or associated with the bottom of a body of water.

Benthos The organisms living on the bottom of an ocean, river, lake, pond, or other body of water.

Biodegradable Pertaining to a substance that can be broken down into simpler components by natural organisms such as bacteria or microbes.

Biodiversity (Biological Diversity) The diversity of life, often divided into three distinct categories: genetic, species, and ecosystem.

Biological oxygen demand The amount of dissolved oxygen needed by aerobic decomposers to break down organic materials in a given volume of water at a certain temperature over a specific period of time. The greater the BOD, the lower the water quality.

Biomass Organic matter produced by photosynthetic producers; total dry weight of all organic matter of plants and animals in an ecosystem.

Biome A major ecological community of organisms, both plant and animal, usually characterized by the dominant vegetation type and climate, such as a tropical rain forest.

Biosphere The large ecosystem encompassing all of the earth's surface, waters, atmosphere, and organisms.

Biota The types of animal and plant life found in an area.

Biotic Living.

Bycatch The portion of a fishery catch consisting of species other than the target species.

Camouflage Coloring, shape, or behavior that allows an organism to blend in with its surroundings.

Carbon dioxide (CO_2) A gas that humans and animals breath out and plants absorb. The burning of fossil fuel and the driving of cars produce large amounts of CO_2.

Carbon monoxide (CO) A gas formed by incomplete combustion of fossil fuels, especially in the internal combustion engine.

Carbon-oxygen cycle The continuous recycling of carbon and oxygen between plants and animals in the ecosystem.

Carcinogen A chemical, ionizing radiation, or virus that causes or promotes the growth of malignant tumors or cancer.

Carnivorous Refers to creatures which eat flesh.

Chemosynthesis The production of energy by bacteria by the chemical oxidation of simple inorganic compounds, such as ammonia or sulfide.

Climate General pattern of atmospheric or weather conditions, seasonal variations, and weather extremes in a region over a long period.

Chlorofluorocarbons (CFCs) A class of chemical compounds containing carbon, fluorine, and chlorine. These chemical compounds, used as aerosol propellants and heat transfer agents in refrigeration/air conditioning systems, are destructive to the stratospheric ozone.

Chlorinated hydrocarbons Organic compounds made up of atoms of carbon, hydrogen, and chlorine. Examples are DDT and PCBs.

Coastal wetland Land along a coastline, extending inland from an estuary that is covered with salt water all or part of the year. Examples are marshes, bays, lagoons, tidal flats, and mangrove swamps.

Coastal zone The relatively warm, nutrient-rich, shallow portion of the ocean that extends from the high-tide mark on land to the edge of the continental shelf.

Commensalism A symbiotic relationship in which one partner benefits from the cohabitation, and the other partner neither benefits nor is harmed.

Commercial fishing Finding and catching fish for sale.

Common-property resource A resource not owned by a particular individual and available for use by everyone, e.g., clean air, fish stocks, migratory birds.

Community A group of organisms of different species which live in a given physical habitat and interact with each other.

Competition Occurs when two or more organisms must work against each other for food or space in order to survive.

Compounds The combination of atoms of two or more different elements held together by the attractive forces called chemical bonds, such as water H_2O, carbon dioxide CO_2, and methane CH_4.

Conservation The careful and organized management and use of a natural resource, emphasizing applied scientific principles.

Consumers Animals and plants that depend on plants directly or indirectly for food.

Contaminate A component not normally present in a specific natural environment. Contamination means not pure. A substance that is contaminated may not necessarily be polluted.

Continental shelf The edges of continental landmasses, now covered with seawater; generally the most productive part of the sea.

Contour strip mining Cutting a series of shelves or terraces along the side of a hill or mountain to remove a mineral such as coal from a deposit found near the earth's surface.

Coriolis force An effect in the earth's rotation that deflects all moving bodies to the right.

"Dead zone" An area within a marine environment in which pollutants have depleted the oxygen supply. Dead zones are devoid of marine life which requires oxygen.

Decomposers Organisms, such as fungi and bacteria, that break down dead plants, animals, and waste products when feeding and return the nutrients contained in them back into the environment.

Deforestation Removal of trees from a forested area without adequate replanting.

Detritus Waste material from dead animals and plants.

Dissolved oxygen The amount of molecular oxygen dissolved in water.

Diversity A measure of the number of different species, along with the number of individuals in each representative species, in a given area.

Dredging A type of surface mining, in which chain buckets and draglines scrape up sand, gravel, and other surface deposits covered with water. Dredging is also used to remove sediment from streams and harbors to maintain shipping channels.

Drift net A gill net suspended vertically from floats at a specific depth, left to float freely.

Ecological diversity The variety of forests, deserts, grasslands, oceans, streams, lakes, and other biological communities interacting with one another and with the non-living environment.

Ecology The scientific study of the interactions of living things and their environment; a community of different species interacting with one another and the chemical and physical factors making up its nonliving environment.

Ecosystem A community of organisms in their environment.

Ecosphere Environment that harbors living organisms.

Ecotone A transitional community between two ecosystems, such as a mangrove which is between terrestrial and marine ecosystems. Ecotones may contain species from both flanking communities as well as species not found in either.

Effluent Wastewater that flows into a receiving stream via a domestic or industrial point source.

Endangered species A species of plant or animal that is present in such small numbers that the species is in danger of disappearing from either all or a significant part of its natural range.

Energy The ability to do work. Plants get energy from the sun and animals get energy from the food they eat. Energy cannot be created or destroyed, only transformed from one state to another.

Energy conservation Reduction or elimination of unnecessary energy and waste.

Environment The aggregate of physical, social, and cultural conditions that influence the life of an individual or community; all external conditions and factors, living and non-living (chemicals and energy), that affect an organism or system during its lifetime.

Environmental assessment A preliminary study to determine the need for an Environmental Impact Statement.

Environmental Impact Statement A report required by the National Environ-mental Policy Act detailing the consequences associated with a proposed major federal action which will significantly affect the environment.

Environmental degradation Depletion or destruction of a potentially renewable resource such as soil, coral, grasslands, forest, wildlife, etc., by using it at a faster rate that it is naturally replenished.

Environmentalists People who are primarily concerned with preventing pollution and degradation of the air, water, soil, and biodiversity of our planet.

Epibenthic Refers to organisms which dwell in the bottommost part of the ocean.

Erosion The wearing away of land surfaces by the action of wind or water.

Estuarine Associated with an estuary.

Estuary An ecosystem in which a river or stream meets ocean waters.

Euphotic zone The uppermost zone of the ocean, into which light penetrates enough to allow plant growth.

Euryhaline Refers to species which can live within a broad range of water salinity.

Eurythermal Refers to species that can live within a broad temperature range.

Eutrophication Inundation of a marine environment with nutrients, leading to excessive growth of phytoplankton, seaweeds, or vascular plants, and often depletion of oxygen.

Evolution Changes in the genetic composition (gene pool) of a population exposed to new environmental conditions as a result of differential reproduction.

Extinction Complete disappearance of a species from the earth. Extinction occurs when a species cannot adapt and successfully reproduce under new environmental conditions.

Fauna A general term for the animal life of an area or region.

Fecal coliform bacteria A type of bacteria found in the colon and waste of warm-blooded mammals, such as humans. Fecal coliform bacteria is considered a contaminant if present in water, food, or beverages.

Federal Register A daily publication of the U.S. government that contains federal administration agency proposed rules, final rules, and other executive branch documents.

Fishery The combination of fish and fishers in a region, fishing for similar or the same species with similar or the same type of gear.

Fish farming Form of aquaculture in which fish are cultivated in a controlled pond or other environment and harvested when they reach the desired size.

Flora A general term for the plant life in an area or region.

Food chain A chain comprised of a producer and consumers. A food chain involves interrelationships among various prey and predators at successive levels within an ecosystem and the passage of energy.

Food web Interconnected food chains showing the feeding relationships in an ecosystem.

Gamete A mature sexual reproductive cell, as a sperm or egg, which unites with another cell to form a new organism.

Genetic diversity Variability in the genetic makeup among individuals within a single species.

Ghost fishing Perpetual entanglement of marine life in fishing lines, nets, and traps purposely or unintentionally discarded in the ocean.

Gill net A net placed vertically in the water to capture fish of a certain size by allowing their heads through the openings but catching their gills in the mesh.

Green house effect The entrapment of solar heat radiating from the earth's surface by exhaust gases generated by the burning of fossil fuels.

Greenhouse gases Gases in the earth's lower atmosphere (troposphere) that cause the greenhouse effect. Examples are carbon dioxide, chloroflourocarbons, ozone, methane, water vapor, and nitrous oxide.

Gyre A circular course or motion.

Habitat The defined space in which an organism, population, or species lives.

Hadley cells Large patterns in which air circulates due to the heat of the sun and the rotation of the earth. Hadley cells generate prevailing winds.

Hazard A substance or action that can cause injury, disease, economic loss, or environmental damage. Hazards come from exposure to various factors in the environment: physical, chemical, biological, and cultural.

Hazardous waste Any discarded chemical that can cause harm by being either flammable, unstable enough to explode, corrosive, or toxic if handled in ways that release them into the environment.

Heavy metals Metallic elements of relatively high molecular weight, such as lead, mercury, arsenic, cadmium, chromium, zinc, etc. Chronic exposure to excessive concentrations of these metals is associated with a variety of health hazards for humans as well as marine wildlife.

Heavy oil Black, high-sulfur, tarlike oil found in deposits of crude oil, tar sands, and oil shale.

Herbivorous Refers to organisms which eat plant matter.

Ichthyologist A zoologist specializing in the study of fishes.

Industrial smog Air pollution consisting mostly of a mixture of sulfur dioxide, suspended droplets of sulfuric acid formed from some of the sulfuric dioxide, and a variety of suspended solid particles; product of burning of fossil fuels (see Photochemical smog).

Intertidal zone An area of coastal land that is covered by water at high tide and uncovered at low tide.

Invertebrates Animals without backbones or internal skeletons

Krill Shrimplike plankton about an inch in length; krill are the primary food of many species of whales in Antarctica.

Lacustrine Associated with lakes.

Landfill An area where solid or solidified waste materials from municipal or industrial sources are buried.

Lake A large natural body of standing fresh water formed when water from precipitation, land runoff, or groundwater flow fills a depression in the earth.

Larva Any animal in an analogous immature form.

Lentic Refers to water which is standing, such as in a lake or pond.

Limiting factor A factor such as temperature, light, water, or a chemical that limits the existence of growth, abundance, or distribution of an organism.

Limnology The study of inland bodies of fresh water and their inhabitants.

Lotic Refers to water which is running, such as in a river or stream.

Mangrove Extremely diverse ecosystems found on tropical shores.

Mariculture The farming of fish in ocean waters.

Marine debris Garbage and trash which is dumped or is carried into the ocean.

Marine Pertaining to the sea.

Marine sanctuary A protected marine area for research and education, where user activities are subject to regulation.

Mass The amount of material in an object.

Maximum sustainable yield The greatest amount of a renewable natural resource that can be removed without diminishing the continuing production and supply of the resource.

Medical waste Discarded materials and chemicals associated with the practice of medicine, such as plastic tubing, needles, syringes, test tubes, and specimen containers.

Metabolism The total chemical activity of the body, or of a part of the body, or of a cell.

Methane A gaseous hydrocarbon that is the main component of natural gas.

Metric ton (t) A unit of mass equal to 1000 kilograms or 2204.62 pounds.

Municipal solid waste Solid waste, including garbage and trash, that originates in households, commercial plants, or construction or demolition sites.

Mutualism A symbiotic relationship in which both partners benefit.

Native species Species that normally live and thrive in a particular ecosystem.

Natural selection The evolutionary process proposed by Charles Darwin. The fundamental principle is that all creatures produce more offspring than their environment can support, and the ones which survive are those best adapted to the environment.

Nautical mile Equal to 6,076 feet, it is the equivalent of 1.15 statute miles.

Nekton Aquatic organisms which swim in the ocean, independent of waves or currents.

Neritic Refers to waters over continental shelves.

Niche An organism's position or place in an ecosystem, based on what it eats, where it lives, and its interactions with other organisms in the ecosystem; total way of life or role of any organism, which includes all physical, chemical, and biological conditions a species needs to live and reproduce in an ecosystem.

Nitrates The form of nitrogen that can be absorbed by plants and animals.

Nonbiodegradable pollutant Material that is not broken down by natural processes, such as plastic.

Non-point source pollution Large or dispersed land areas such as crop fields, streets, and lawns that discharge pollutants into the environment over a large area.

Non-renewable resources Resources existing in fixed stock which have the potential for renewing only by geological, physical, and chemical processes taking place over hundred of millions of years, e.g., coal, copper, oil.

Nutrients Elements which are necessary for an organism to live, grow, and reproduce.

Oceanic Relating to waters over ocean deeps.

Oceanography The study of the physics, chemistry, biology, and geology of the ocean.

Open sea The part of the ocean that is beyond the continental shelf.

Organic Any molecule in which carbon is predominant.

Oxygen (O_2) A gas in the air given off by plants during photosynthesis and breathed in by animals and people during respiration.

Ozone (O_3) A form of oxygen formed in the stratosphere by interaction between discharge of electricity and oxygen. Ozone protects organisms on earth from the sun's intense ultraviolet radiation. In the troposphere, ozone is an irritant found in smog. Ozone is used commercially for sterilizing, bleaching, and purifying.

Ozonosphere A layer of naturally occurring oxygen compound located about 20 miles above the earth, which acts as an ultraviolet radiation filter and atmospheric insulator.

Palustrine Associated with a marsh, swamp, or bog.

Parasitism A relationship between two disparate species, in which the host is harmed, killed, or impoverished.

Pelagic Living afloat in the water column or open sea.

Photochemical smog A complex mixture of air pollutants produced in the atmosphere by the reaction of hydrocarbons and nitrogen oxides under the influence of sunlight. These pollutants are   produced primarily from combustion engines.

Photosynthesis The process by which plants combine carbon dioxide from the air, energy from the sun, and chlorophyll in their leaves, to produce food (sugars) for growth and release oxygen back into the environment.

Phytoplankton Plankton composed of plants, which form the base of the marine food chain.

Plankton Tiny plants and animals that drift in the surface waters. Plankton provide the base of the food chain in the sea.

Point source pollution A single identifiable source that discharges pollutants into the environment, such as an industrial plant, drainpipes, smoke stacks, and car exhaust pipes.

Pollutant Something that can adversely affect organisms or materials. Something is polluted if it is rendered not suitable for its intended use.

Pollution The addition of one or more chemical or physical agents to the environment in an amount, at a rate, and/or in a location that threatens the health or normal function of the species in that environment.

Polychlorinated biphenyls (PCBs) Poisonous chemicals used to cool machinery.

Population A group of individuals of the same species that live in a given area.

PPB Parts per billion. A measure of a concentration of a substance (usually a toxic material) in liquid.

Precipitation Water that falls to the earth in the form of rain, snow, sleet, or hail.

Predator-prey relationship Predators are usually considered animals that kill and eat other animals (prey). This relationship helps to keep the number of organisms in an ecosystem in balance with their environment. However, biologists commonly refer to herbivorous animals as "predators" of plants.

Primary consumers Animals that eat plants and make up the second level of the food pyramid.

Primary productivity The rate at which plants change energy of the sun into more generally usable forms (i.e., food).

Producer An organism that uses solar energy (green plants) or chemical energy (some bacteria) to manufacture the organic compounds it needs as nutrients from simple inorganic compounds obtained from its environment. Producers are the bottom of the food pyramid.

Purse seine Large nets used in commercial fisheries which encircle and enclose schools of fish.

Radioactive waste Radioactive waste products of nuclear power plants, research, medicine, weapons production, or other processes involving nuclear reactions.

Raw sewage Wastewater that has not undergone any treatment for the removal of pollutants.

Recycling Collecting and reprocessing a resource so it can be made into new products. Recycled materials include plastic, glass, paper, and aluminum.

Red tide Reddish-brown discoloration of the water caused by algae blooms, believed to be caused by pollution. Red tide causes the death of marine biota and the accumulation of toxins in mussels and clams, making them hazardous for human consumption.

Recruitment The addition of members to a population.

Renewable resource A resource that can last indefinitely because it can be replaced rapidly by natural processes, e.g., trees, grassland, livestock, surface and groundwater. Renewable resources can be depleted if used at a rate above the renewal rate.

Respiration The physical and chemical processes in an organism by which oxygen and carbohydrates are assimilated into the system and the oxidation products, carbon dioxide and water are given off.

Riverine Associated with rivers.

Runoff Fresh water from precipitation and melting ice that flows on the earth's surface into nearby rivers, streams, lakes, wetlands, and reservoirs.

Salt marshes Tidal wetlands found in temperate zones.

Secondary consumers Animals that eat other animals and make up the third trophic level in the food chain.

Sediment Soil particles, sand, clay, or other substances that settle to the bottom of a body of water.

Sewage Wastewater from homes, businesses, or industries; mainly refers to the water transport of cooking, cleaning, or bathroom waste.

Sewage treatment plant A facility designed to receive the wastewater from domestic sources and to remove materials that damage water quality and threaten public health when discharged into receiving streams.

Sludge A thick, mud-like mixture of toxic chemicals, infectious agents, and suspended particulate matter removed from wastewater at sewage treatment plants.

Smog See Photochemical smog and Industrial smog.

Species A group of organisms that resemble one another in appearance, behavior, chemical makeup and processes, and genetic structure. Organisms that reproduce sexually are classified as members of the same species only if they can actually or potentially interbreed with one another and produce fertile offspring.

Species density The total number of individuals of a specific species found in a specific area of a habitat for a specified time period.

Species diversity The number of different species and their relative abundance in a given area.

Sport fishing Finding and catching fish for recreation.

Stenothermal Living within a narrow temperature range.

Stocks In ecological cycles and models, the amounts of a material in a certain medium.

Stratosphere The second layer of the atmosphere, extending from about eleven to thirty miles above the earth's surface. It contains small amounts of gaseous ozone which filters out about ninety-nine percent of the incoming harmful ultraviolet radiation emitted by the sun.

Stressed waters A portion of an aquatic environment with poor species diversity due to human actions.

Strip mining Form of surface mining in which bulldozers, power shovels, or stripping wheels remove large chunks of the earth's surface in strips.

Substrate (substratum) The base or material on which an organism lives; the solid surface on which an organism moves about or attaches, such as sediment, rock, or sand on the bottom of the ocean.

Succession, ecological A process in which communities of plants and animals in a particular area are replaced over time by a series of different and usually more complex communities.

Sustainable (economic) development Forms of economic growth and activities that do not deplete or degrade natural resources upon which present and future economic growth depend.

Sustainable yield The highest rate at which a potentially renewable resource can be used without reducing its available supply; the optimum annual catch that can be derived indefinitely from harvested species, without causing a stock failure.

Symbiosis A harmonious relationship between two disparate species.

System A regular interacting or interdependent group of items forming a unified whole.

Tar balls Nonvolatile hydrocarbon clumps remaining in the water after crude oil is spilled or discharged into the marine environment. Tar balls can be fatal to sea birds, as they coat the birds' feathers, causing them to lose their insulative protection and their ability to fly to forage for food.

Terrestrial Pertaining to land.

Toxicity The ability of a chemical to cause injury to an organism by itself or by a substance formed when it is taken into a body. Toxicity is not a property of the substance itself.

Toxins Poisonous substances; chemicals that can cause injury to an organism.

Tragedy of the commons Depletion or degradation of a resource to which people have free and unmanaged access. An example is the depletion of commercially desirable species of fish in the open ocean beyond areas controlled by coastal countries.

Trawl A funnel-shaped fishing net towed behind a fishing vessel.

Trophic level The level of a species in the food chain. The farther removed the species is from primary production, the higher its trophic level.

Troposphere The innermost layer of the atmosphere. It contains about 95% of the mass of the earth's air and extends about 11 miles above sea level.

Turbidity A state of reduced water clarity caused by suspended particles.

Upwelling A process whereby deeper, cold nutrient-rich waters rise up to replace warmer surface waters.

Virus The simplest kind of living organism, viruses live as parasites within the cells of bacteria and are the principal agents of disease.

VOCs Volatile Organic Compounds. Among the most commonly found contaminants in groundwater, including vinyl chloride (chlorethylene) and trichlorethylene (TCE).

Vertebrates Animals possessing backbones and internal skeletons.

Wastewater Water discharged from homes, businesses, and industries that contains dissolved, suspended, and particulate organic or inorganic material. The term is also used as a synonym for sewage.

Wastewater lagoon A large pond where air, sunlight, and microorganisms break down wastes, allowing solids to settle out, and killing some disease-causing bacteria. Water typically remains in a lagoon for thirty days, then it is treated with chlorine and pumped out for use by a city or spread over agricultural fields.

Water cycle Describes the continuous recycling of water in an ecosystem.

Water pollution Any physical or chemical change in surface water or groundwater that can harm living organisms or make water unfit for certain uses.

Watershed A land area that delivers the water, sediment, and dissolved substances via small streams to a major stream or river.

Water treatment The processing of surface water or well water for distribution in a public drinking system.

Wetlands Areas where water is the primary factor controlling the environment and the associated plant and animal life, excluding streams, lakes, and the open ocean.

Wildlife All free, undomesticated species.

Zooplankton The small, often microscopic animals in the aquatic environment that drift along with the currents.

Zooxanthellae Microscopic dinoflagellate algae that live in a mutualistic relationship in the tissues of certain marine invertebrates, such as reef-building corals.

APPENDIX 1

MARINE CONSERVATION LEGISLATION

The following are the major U.S. laws which directly or indirectly protect the marine environment. They have been divided into three general categories: marine and coastal protected areas, pollution and hazardous substances, and species protection.

MARINE AND COASTAL PROTECTED AREAS

NOAA National Marine Sanctuary Program

In response to growing threats to fragile marine ecosystems, the National Oceanic and Atmospheric Administration (NOAA) created the national Marine Sanctuary Program (NMSP) in 1972. (See Chapter 11.) This program was designed to promote–for research, education, management, and preservation–national areas of historic, scientific and ecological significance. The NOAA NMSP consists of fourteen sanctuaries, ranging in size from less than one nautical mile to several thousand nautical miles. Current sanctuaries include the Channel Islands (off the Southern California coast), Cordell Bank (twenty miles west of Point Reyes, CA), Fagatele Bay (American Samoa), Florida Keys, the Flower Gardens Banks (in the Gulf of Mexico, 120 miles from the LA-Texas border), Gray's Reef (Georgia), Gulf of the Farallones (NW of San Francisco), Hawaiian Islands, Humpback Whale, Key Largo, Looe Key (Florida), Monterey Bay, Olympic Coast (along the northern Pacific coast of Washington state), Stellwagen Bank (off Provincetown, MA), and the *Monitor* (off N. Carolina). NOAA continually researches and recommends to the U.S. Congress new sites to be designated as sanctuaries. Sites currently under observation include Northwest straits (North of Seattle, WA) and Norfolk Canyon (off the coasts of VA and N. Carolina), and Thunder Bay (Michigan).

Coastal Zone Management Act

Adopted by Congress in 1972, the Coastal Zone Management Act provides federal grants to states to develop coastal zone management plans that balance the pressure for economic development with the need for environmental protection. With support from NOAA, management policies for the coastal zone are ideally intended to protect coastal natural resources (including estuaries, bays, beaches, and fish and wildlife and their habitats) and to encourage management plans for estuaries, bays and harbors. As of 1985, twenty-eight states have developed federally approved plans.

Coastal Barrier Resources Act
The Coastal Barrier Resources Act seeks to protect coastal barriers in the Atlantic Ocean and Gulf of Mexico by discouraging development. The act prohibits expenditures of federal money for development of infrastructure such as roads, sewer systems, water supply systems, bridges, and jetties. The act established the Coastal Barrier Resources System (CBRS), which is composed of thousands of miles of barrier islands along the Atlantic and Gulf coasts.

MAJOR POLLUTION AND HAZARDOUS SUBSTANCES LAWS

National Environmental Policy Act (NEPA)
The National Environmental Policy Act is perhaps the shortest, yet one of the most important, of all the federal environmental laws in this country. The law directs the federal government and federal agencies to take into consideration the environmental impact of all its actions. This requirement is considered the origin of what is commonly known as an Environmental Impact Statement (EIS). The idea of NEPA was to require that certain procedural steps be taken by an agency prior to the initiation of any project to assure that the decision maker and the public would be apprised of the environmental consequences of the project.

Clean Air Act (CAA)
Because airborne pollutants are a threat to the marine as well as terrestrial environments, the Clean Air Act has far-reaching effects. This key legislation, which is several hundred pages long, was first enacted in 1955 and was rewritten in 1963, 1965, and 1967, and 1970, then significantly amended in 1977. The Clean Air Act is based on National Ambient Air Quality Standards, which establish maximum allowable concentrations for the pollutants that the EPA (Environmental Protection Agency) deems to be particularly widespread and dangerous. Pollutants identified by the EPA include carbon monoxide, sulfur dioxide, particulate matter, ozone, nitrogen oxide, and lead. Although the EPA sets the nationally acceptable air quality standards, it is up to the states to comply with them. Each state has been charged with developing a State Implementation Plan setting out its various pollution control programs and requirements by a specific deadline (all of which have expired) set by Congress.

The Clean Air Act also imposes performance standards on factories, power plants, and other "stationary sources" of air pollution, particularly automobile emissions. Recognizing that motor vehicles were responsible for an increasingly large share of air pollution, in 1970, Congress demanded that all new cars be designed to emit far less carbon monoxide, hydrocarbons, and nitrogen oxides. In order to comply with this statute, car manufacturers were forced to develop and commercialize new technologies.

Although the Clean Air act has helped regulate air quality in general, there are still widespread violations in the allowable concentrations of carbon monoxide, ozone, and particulate matter in many areas. Nor has the Clean Air Act controlled acid rain.

Clean Water Act (CWA)

Passed by Congress in 1972, this is the primary federal law governing the discharge of pollutants in inland and coastal waterways; it is administered by the EPA. The primary objective of the Clean Water Act was to maintain and restore the chemical, physical, and biological integrity of U.S. waters. The 1972 act set a goal of fishable, swimmable waters nationwide by 1983. Congress also anticipated (but did not require) an end to all discharges of pollutants into waterways by 1985. To accomplish this, Congress established a combined federal and state system of controls to implement clean water programs. The CWA consists of two major parts: (1) the federal grant program to help municipalities build sewage treatment plants; and (2) the pollution control programs, which consist of regulatory requirements and permits for industrial and municipal dischargers. To help implement these goals, the government initiated a $5 billion-a-year federal grants program to finance construction of local sewage treatment systems.

Unfortunately, many of the goals of the Clean Water Act have not been met. One of the most problematic factors is that roughly half of all water pollution does not come from specific sources of discharge (point pollution), but from non-point sources, and the legislation does not apply to non-point sources. Another problem with the CWA is that is grants permits for dredged or fill material to be put into "waters of the United States," if a developer can show that no practical alternative is available that would not involve filling in waters. The term "waters of the United States" has been criticized as being too vague, because it does not protect inland and coastal wetlands from being targeted for development. As a result, there have been heated political battles for permits granted under the provisions of the CWA for shopping malls, marinas, and housing developments in these valuable wetlands. Also, see page 114 on controversy over reauthorization of the clean water act.

Safe Drinking Water Act (SDWA)
The SDWA, which was passed in 1974 and amended in 1986, requires the EPA (Environmental Protection Agency) to set standards, known as Maximum Contaminant Levels for contaminants in public drinking water supplies. Each state is charged with enforcing this law. The SDWA also seeks to protect underground drinking water supplies. The EPA sets standards and issues permits for underground disposal of liquid wastes, and it offers federal funding and support for state programs to protect underground water supplies. Also, see page 114 on controversy over reauthorization of the clean water act.

Ocean Dumping Act
In 1988, the Untied States enacted this legislation mandating a phase out of ocean dumping of industrial waste and sewage sludge. The Ocean Dumping Act (Title I of the Marine Protection Research and Sanctuaries Act) governs the disposal of all materials into the ocean, including sewage sludge, industrial waste, and dredged materials. It forbids outright the ocean dumping of radiological, chemical, and biological warfare agents and high-level radioactive waste and requires an EPA permit to dump other materials at sea. The EPA and the Army Corps of Engineers are the permitting agencies for ocean dumping, while NOAA is responsible for research and monitoring and the U.S. Coast Guard for surveillance and enforcement. A recent amendment to the Ocean Dumping Act required dumpers either to end ocean dumping by 1992 or pay escalating permit fees as long as dumping continues.

Marine Plastics Pollution Research and Control Act
Enacted in 1987, the Marine Plastics Pollution Research and Control Act bans the dumping of plastics, including synthetic fishing nets, in U.S. waterways and coastal zones and by U.S. vessels anywhere in the ocean. Under MARPOL Annex V, (see International Treaties and Laws section) the U.S. and thirty-eight other signatory countries have agreed to ban the dumping of plastics by vessels. The act also requires several studies to be conducted by the EPA and NOAA to determine the extent of the impacts of plastics pollution on fisheries and wildlife and to explore methods to reduce such waste in the marine environment. This act applies to all water craft, including small recreational vessels.

The Port and Tanker Safety Act of 1978

This law empowers the U.S. Coast Guard to supervise vessel and port operations and to set standards for the handling of dangerous substances. The act and the regulations cover the design, construction, alteration, repair, maintenance, operation, equipping, personnel, and manning of vessels and set minimum standards for ballast tanks, oil-washing systems, and cargo-protection systems. The act also mandates a national program for annual inspection of vessels.

The Intervention on the High Seas Act

This law authorizes the U.S. Coast Guard to take measures on the high seas to prevent, mitigate, or eliminate the danger of harm from any oil spill on the high seas that poses "a grave and imminent danger to the coastline or related interests of the United States." This "danger" can include threats to human health, fish, marine resources, and wildlife.

Oil Pollution Act of 1990

Spurred by the catastrophic oil spill of the *Exxon Valdez* in the pristine Prince William Sound coastal area of Alaska and a rapid succession of other oil spills, Congress passed this comprehensive oil spill liability legislation in 1990. The Oil Pollution Act defines stiff terms of cleanup and damage assessment and rapid repayment of damages to those incurring loss or injury from an oil spill. The law requires that new oil tankers be built with double hulls and that all existing single-hull vessels be phased out by the year 2014. The law also increases federal liability limits for vessels from $150 per gross ton to $1200 per gross ton per spill and creates a $1 billion federal oil spill fund, which is financed by the oil industry. The law also requires review of licensed seamen's driver's license records along with drug and alcohol testing, establishes ten regional oil spill response groups, and requires federally approved oil spill contingency plans for vessels and oil facilities.

Minerals Management Service (MMS)

The U.S. Minerals Management Service is the lead regulatory agency over offshore oil and gas activities in federal waters, and hence the regulators for handling and treating garbage generated during the course of these activities. These regulatory requirements are set out in a pollution prevention and control order that restricts the disposal of solid waste materials into the ocean. Personnel who oversee these activities conduct routine inspections of offshore operations to determine if such laws are followed.

Comprehensive Environmental Response Compensation and Liability Act (SUPERFUND)

The CERCLA, or Superfund, enacted in 1980 and amended in 1986, provides emergency response and cleanup provisions for chemical spills and for releases from hazardous waste facilities that threaten human health or the environment. The program has been used to identify numerous existing hazardous waste sites, many in coastal or underwater sites, that need to be cleaned up. Any site where a hazardous substance has been released is a potential Superfund site. As of 1989, the EPA had identified more than 26,000 Superfund sites. CERCLA assigns liability and financial responsibility for cleanups to many parties involved with the release of hazardous substances. If no solvent private party can be found to foot the bill to cleanup a Superfund site, the Superfund is available.

The Nuclear Waste Policy Act of 1980

The law places responsibility for disposal of low-level radioactive waste on the state in which the waste was produced. The act also encourages development of regional, multi-state compacts. There are presently nine such regional compacts, including all states except California and Texas.

The Nuclear Waste Policy Act of 1982

This legislation sought to solve the problem of high-level radioactive waste, primarily the spent reactor fuel now in temporary storage around the country. It charges the Department of Energy with developing a permanent high-level waste depository by 1998. DOE must consult with state governments on the siting of a nuclear-waste dump within its borders. A state can veto DOE's plans, but the state's veto can be overruled by Congress.

SPECIES PRESERVATION

Endangered Species Act

The Endangered Species Act (ESA) is the most comprehensive federal law for the protection of species diversity and species habitats. The law gives the Secretary of the Interior, acting through the U.S. Fish and Wildlife Service (FWS), responsibility for the recovery of terrestrial species and some marine species, and the Secretary of Commerce, through the National Marine Fisheries Service (NMFS), responsibility for the recovery of most marine species. The ESA authorizes the Secretaries to identify endangered or threatened species, designate habitats critical to their survival, establish and conduct programs for their recovery, enter into agreements with states, and assist other countries to conserve endangered and threatened species. The federal government is also empowered to enforce prohibitions against or issue permits controlling the taking of and trading in endangered and threatened species. Under the ESA, federal agencies are prohibited from funding, authorizing, or carrying out any projects that jeopardize the existence of or modify the habitats of endangered species.

Marine Mammal Protection Act (MMPA)

The MMPA was enacted in 1972 to protect all marine mammal species, many of which were in danger of extinction or depletion as a result of human activities. The National Marine Fisheries Service manages all cetaceans (whales and dolphins) and all pinnipeds except walruses, (i.e., seals and sea lions). The U.S. Fish and Wildlife Service manages polar bears, walruses, sea otters, manatees, and dugongs. The MMPA primarily prohibits hunting, wounding, and harassing marine mammals, and prohibits importation of marine mammals or their products.

Marine Entanglement Research Program (MERP)

In 1984, Congress recognized the problems of marine debris and entanglement and directed the National Marine Fisheries Service (NMFS) to develop a research program in consultation and concurrence with the Marine Mammal Commission. The MERP conducts educational activities aimed at debris generators, oversees the operation of two marine debris information offices, sponsors research on the origin, amount, distribution, and fate of marine debris, and explores ways to reduce the amount of nonbiodegradable material lost or disposed of at sea.

Fisheries Conservation and Management Act (FCMA)

Enacted in 1976 in response to the increasing presence of foreign fleets which were depleting marine species, the FCMA enabled the U.S. to extend its jurisdiction and control over all marine fisheries resources within 200 miles of the U.S. coast. The FCMA established eight regional fishery management councils composed of state and federal fishery officials and industry representatives. The councils are charged with preparing, monitoring, and revising fishery management plans. The plans must include measures to rebuild and restore fish stocks, prevent overfishing, and assure an optimum yield from each fishery. Fishery management plans, which are approved by the Secretary of Commerce acting through the NMFS, may include seasonal restrictions, gear restrictions, size limitations, and limited entry to the fishery. Coastal states maintain management control over fishery resources within state waters (three miles offshore for all states except Florida and Texas, where it is nine miles).

APPENDIX 2
FEDERAL ENVIRONMENTAL AGENCIES
AND PROGRAMS

Appendices 2 and 3 contain the addresses of U.S. Federal departments, agencies, and offices, and non-governmental environmental organizations (NGOs). These lists are by no means exhaustive. Many of these agencies and organizations have regional offices, which can be contacted by consulting your phone directory or local library, or by writing to the main offices listed in these appendixes.

Department of Commerce
National Oceanic and Atmospheric Administration (NOAA)
National Marine Sanctuaries Program
1305 East-West Highway, Bldg. 4, 12th FL
Silver Springs, MD 20910
(301) 713-3125
Jurisdiction: national marine sanctuaries

NOAA Marine Debris Information Office
c/o Center for Marine Conservation
1725 DeSales Street, NW
Washington, DC 20036
(202) 429-5609

National Marine Fisheries Service
Office of Protected Resources
1335 East-West Highway, Bldg. 1
Silver Springs, MD 20910
(301) 713-2239
Jurisdiction: endangered species, marine mammals, marine habitats

Office of Ocean and Coastal Resource Management
National Ocean Service
1825 Connecticut Avenue, NW
Washington, DC 20235
(202) 606-4111
Jurisdiction: coastal zone management

Department of Defense
Office of Public Affairs–Army Corps of Engineers
20 Massachusetts Avenue, NW
Washington, DC 20314-1000
(202) 272-0011
Jurisdiction: wetlands dredge and fill

Department of the Interior
Division of Endangered Species
U.S. Fish and Wildlife Service
4401 N. Fairfax Drive, Room 452
Arlington, VA 22203
(703) 358-2171
Jurisdiction: endangered species

Minerals Management Service
1849 C Street, NW
Washington, DC 20240
(202) 208-3983
Jurisdiction: offshore oil and gas development

Department of Transportation
United States Coast Guard
2100 2nd Street, SW
Washington, DC 20593
(202) 267-2229
Jurisdiction: marine debris violations (call for local Coast Guard office), oil and hazardous materials transport

Environmental Protection Agency
Public Liaisons Division
401 M Street, SW, A107
Washington, DC 20460
(202) 260-4361
Jurisdiction: general information on EPA programs

APPENDIX 3

NON-GOVERNMENTAL
ENVIRONMENTAL ORGANIZATIONS

There are numerous non-governmental organizations (NGOs) which focus on environmental issues. The following is a list of organizations whose goals include marine conservation. For a complete directory of NGOs in the United States, *The Directory of National Environmental Organizations*, edited by John C. Brainard (Roger N. McGrath Publishers, St. Paul, MI) lists over 675 NGOs alphabetically, by subject, and by geographic location.

Alliance for Environmental Education
Serves as an advocate for environmental conservation through education and advanced communication, cooperation, and exchange among organizations. Has established a network of interactive environmental education centers based in colleges, universities, and institutions throughout the U.S.
PO Box 368
The Plains, VA 22171
(703) 253-5812

American Association of Zoological Parks and Aquariums
A professional organization which represents 160 accredited zoos and aquariums in the U.S. The primary goal is to further wildlife conservation and education and to enforce a code of ethics for all individual members and zoological institutions.
7970-D Old Georgetown Rd.
Bethesda, MD 20814
(301) 907-7777

American Cetacean Society
Created in 1967, ACS is a California-based non-profit education organization with international membership. The goals of ACS are legislative action and informing the public about cetaceans and the need for conservation. Offers its members programs and teaching materials, such as The *Whale Watchers* packet and the *Spyhopper* newsletter. Whale teaching kits are available for professional educators.

PO Box 2639
San Pedro, CA 90731-0943
(310) 548-6279

American Geographical Society
An organization whose purpose is to expand and disseminate geographical knowledge through publications, awards, travel programs, lectures, and consulting, with an emphasis on ecology and environmental issues abroad and in the U.S.
156 Fifth Avenue, Suite 600
New York, NY 10010-7002
(212) 242-0214

American Littoral Society
A national, non-profit public interest organization of professional and amateur naturalists. Their goal is to encourage a better understanding of aquatic environments and provide a unified voice advocating conservation. The shore and adjacent wetlands, bays, rivers–the littoral zone–is the special area of interest and concern for members of the American Littoral Society. In addition to sponsoring conservation legislation, ALS conducts trips and publishes *Underwater Naturalist.*
Sandy Hook
Highlands, NJ 07732
(908) 291-0055

American Oceans Campaign
Non-profit organization which focuses on public education and beach and underwater cleanups. President is actor Ted Danson, and AOC has strong ties to pro-environment Hollywood celebrities. AOC's Washington, DC office focuses on legislation such as the Clean Water Act and Safe Drinking Water Act and lobbies on issues such as water pollution issues, debris from ships, offshore oil drilling and tanker safety, and national and international fisheries management. AOC publishes a quarterly newsletter, *Splash.*

AOC California Office
725 Arizona Ave., Suite # 102
Santa Monica, CA 90401
(310) 576-6162

AOC Legislative Branch
235 Pennsylvania Avenue, SE
Washington, DC 20003
(202) 544-3526

American Rivers
Leading conservation organization for protection and restoration of U.S. river systems. Has effectively preserved over 10,000 river miles

for clean water, endangered fish and wildlife, recreation, and aesthetic beauty.
801 Pennsylvania Ave. SE, Suite 400
Washington, DC 20003
(202) 547-6900

California Marine Mammal Center
A non-profit organization established in 1975 to rescue and rehabilitate sick, injured, or distressed marine mammals. CMMC has a worldwide membership, and its work is carried out by 300 volunteers and a team of experts.
Marine Headlands
Golden Gate National Recreation Area
Sausalito, CA 94965
(415) 331-SEAL

Caribbean Conservation Corporation
A non-profit organization, established in 1969, which is dedicated to the conservation and preservation of sea turtles and other coastal and marine wildlife through protection of natural areas, research, advocacy, education, and training. CCC publishes a quarterly newsletter, *Velador*, disseminates materials, and makes environmental presentations.
P.O. Box 2866
Gainsville, FL 32602
(904) 373-6441 (800) 678-7853

CEDAM
A membership organization which specializes in diving and environmental awareness expeditions led by noted scientists and naturalists.
Fox Road
Croton-On-Hudson, NY 10520
(914) 271-5365

Center for Environmental Education
Provides comprehensive and up-to-date information on environmental issues. Offers publications, educational programs, and information services.
46 Prince Street
Rochester, NY 14607
(716) 271-3550

Center for Marine Conservation
The leading non-profit organization dedicated to the conservation
of marine wildlife and habitats focuses on five major goals: conserving marine habitats; preventing marine pollution; fisheries conservation; protecting endangered species; and providing educational
materials and programs on biodiversity. Publishes *Marine
Conservation, Sanctuary Currents,* and annual cleanup reports
(Coastal Connection).
1725 DeSales Street, NW, Suite 500
Washington, DC 20236
(202) 429-5609

Citizen's Clearinghouse for Hazardous Waste
Assists communities to fight environmental threats through grass-
roots efforts.
PO Box 6806
Falls Church, VA 2240
(703) 237-2249

Clean Ocean Action
A non-profit organization focusing on cleaning up and protecting
the waters off the New York/New Jersey coast. A strong and outspoken advocate against dredging and non-point source pollution.
Publishes a monthly newsletter, *Ocean Advocate,* conducts beach
cleanups twice a year, and disseminates environmental educational
materials.
PO Box 505
Highlands, NJ 07732
(908) 872-0111

Clean Water Action Project
National citizens' organization working for clean and safe water,
control of toxic chemicals, protection and conservation of wetlands,
groundwater and coastal waters, safe solid waste management,
public health, and human environmental safety.
1320 18th St. NW, 3rd Floor
Washington, DC 20036
(202) 457-1286

Coast Alliance
A non-profit educational organization concerned with coastal
development issues on all four U.S. coasts. Its four main focuses are
the National Flood Insurance Program, the Coastal Barrier
Resources Act, contaminated sediments, and coastal zone reautho-

rization. Coast Alliance hosts workshops for public education and disseminates educational materials.
235 Pennsylvania Avenue, SE
Washington DC 20003
(202) 546-9554

Coastal Conservation Association

A non-profit organization emphasizing conservation of coastal and marine wildlife and habitats. Sponsors member events and publishes a bi-monthly magazine, *Tides.*
4801 Woodway, Suite 220
Houston, TX 77056
(713) 626-4222

Coastal Society

A non-profit organization of individuals affiliated with coastal resource management agencies, colleges and universities, marine advisory groups, and local advocacy groups. Coastal Society seeks to be an umbrella for communication and cooperation among all individuals who interact with the coastal environment. The Society, whose emphasis is on sustainable development, encompasses the entire U.S. coastline and Great Lakes, as well as having a small international membership in Japan and Europe. The Coastal Society holds a conference every two years which is open to the public and publishes a quarterly newsletter, *Bulletin.*
PO Box 25408
Alexandria, VA 22313-5408
(703) 768-1599

Conservation Law Foundation of New England

This organization, established in 1966, is comprised of attorneys, scientists, and policy specialists who use the law in the interest of resource management, environmental protection, and public health in the new England area. Among its achievements, CLF successfully blocked three attempts by the U.S. government to drill for oil and gas on George's Bank, the world's most productive fishery.
3 Joy Street
Boston, MA 02108-1497
(617) 742-2540

Conservation International

Dedicated to the conservation of ecosystems and biological diversity and the ecological processes that support life on earth. CI has

four major themes: focusing on entire ecosystems; integrating economic interests with ecological interests; creating a base of scientific knowledge necessary to make conservation-minded decisions; and making it possible for conservation to be understood and implemented at the local level.
1015 18th St., NW, Suite 1000
Washington, DC 20036
(202) 429-5660

The Coral Reef Alliance (C.O.R.A.L.)
A comprehensive non-profit program which promotes coral reef conservation by enlisting the support of the scuba diving community. C.O.R.A.L. disseminates information via newsletters, brochures, meetings, and electronic bulletin boards on current conditions in the Florida reefs, threats to the reefs, and recommended solutions. Through diving industry donations, C.O.R.A.L. plans to provide long-term support for NGOs working on coral reef conservation, particularly grass roots efforts that have little or no access to other forms of financial support.
809 Delaware Street
Berkeley, CA 94710
(510) 528-2492

Cousteau Society
Dedicated to the protection and improvement of the quality of life. The Society produces television films, books, membership publications and articles, and offers lectures and a summer field study program. Publishes newsletter, *Calypso Log*.
930 West 21 Street
Norfolk, VA 23517
(804) 627-1144

Defenders of Wildlife
A national non-profit organization which utilizes public education, litigation, and advocacy of progressive public policies to protect the diversity of wildlife and preserve the habitat critical to its survival.
1244 19 Street, NW
Washington, DC 20036
(202) 659-9510

Earth Island Institute
This organization develops innovative projects for the conservation, preservation, and restoration of the global environment.

300 Broadway, Suite 28
San Francisco, CA 94133
(415) 788-3666

Earthwatch

An organization which sends volunteers to work with scientists around the world who are saving terrestrial and marine habitats and endangered species, preserving archeological finds, and studying the effects of pollution.
PO Box 403N
680 Mt. Auburn St.
Watertown, MA 02172
(800) 776-0188

Ecological Society of America (ESA)

A nonpartisan, non-profit, scientific society founded in 1915 to encourage the scientific study of the interrelations of organisms and their environments. The Society's 7,400 members in the U.S., Canada, Mexico, and 62 other nations conduct research, teach, and aid decision-makers in universities, government agencies, industry, and conservation organizations. ESA strives to stimulate and publish research on the interrelations of organisms and their environment, facilitate an exchange of ideas among those interested in ecology, and instill ecological principles in the decision-making of society at large. Through its Public Affairs Office, ESA maintains a computerized data bank of more than 3,000 ESA members who can provide expert scientific information to Congress and other groups on issues affecting domestic and international environmental quality. The Network provides rapid answers to questions about effects of human activities on animals, plants, and microorganisms in both natural and managed ecosystems. ESA publishes the bimonthly *Ecology* and the quarterly *Ecological Monographs* and *Ecological Applications,* the quarterly *Bulletin of the Ecological Society of America,* and the *Newsletter of the Ecological Society of America.*
2010 Massachusetts Ave. NW, Suite 420
Washington, DC 20036
(202) 833-8773

Environmental Action

Environmental Action was founded in 1970 by the founders of the first Earth Day. EA is a political action organization that works to protect the earth's environment through pollution prevention. EA's targets are solid waste, toxic substances, drinking water, recycling, global warming, plutonium production, utility policy, ozone depletion, and acid rain.

1525 New Hampshire Avenue, NW	6930 Carroll Ave., Suite 600
Washington, DC 20036	Takoma Park, MD 20912
(202) 745-4870	(301) 891-1100

Environmental Data Research Institute, Inc.

EDRI maintains a large database on environmental grants and has published a 490-page comprehensive directory. EDRI provides the environmental community with information on funding.
797 Elmwood Ave.
Rochester, NY 14620
(716) 473-3090

Environmental Defense Fund

Combines outstanding scientific research and legal action on a wide range of subjects including pesticides, recycling, renewable energy, radon, and toxic waste.
257 Park Avenue South
New York, NY 10010
(212) 505-2100

The Environmental Exchange

Aids grassroots environmental projects by providing information to the public about environmental projects.
1930 18th St. NW, Suite 24
Washington, DC 20009
(202) 387-2182

Environmental Support Center

This center operates programs to strengthen regional, state, local and grassroots organizations working on environmental issues. ESC pays most of the cost of contracting with professionals to provide training and technical assistance to those groups in fundraising, organizational development, and strategic planning.
1875 Connecticut Ave, NW, Suite 340
Washington, DC 20009
(202) 328-7813

Friends of the Earth/Oceanic Society

A national and international organization which is active in fighting for many world issues including rain forest and marine environment protection, preservation and restoration of ecosystems, and renewable energy development.

218 D Street, SE
Washington, DC 20003
(202) 544-2600

Greenpeace USA

National and international direct action and lobbying on rain forests, toxic wastes, ocean and air pollution, whales, nuclear radiation, and other global issues. Greenpeace activities often make world headlines because of its members' direct confrontation with governments and industrial corporations. Publishes a bimonthly magazine, *Greenpeace.*
1436 U Street, NW
Washington, DC 20009
(202) 426-1177

Inform

An organization for environmental research and education that identifies and reports on practical actions for the preservation and conservation of natural resources and public health. This group publishes six reports a year on such issues as hazardous waste reduction, air quality, and land and water conservation.
381 Park Ave. S.
New York., NY 10016
(212) 689-4040

International Oceanographic Society

A non-profit marine education and science issues clearinghouse for teachers, students, and members. IOS is located on the campus of the Rosenstiel School of Marine and Atmospheric Sciences, and members have access to the college library and faculty. IOS publishes the quarterly *Sea Frontiers,* the book *Ocean Life,* as well as numerous pamphlets about training in careers in marine science.
4600 Rickenbacker Causeway
Virginia Key, Miami, FL 33149
(305) 361-4888

International Union for the Conservation of Nature

A non-governmental agency of international scope that promotes measures to conserve wildlife and natural resources.
Avenue de Mont Blanc
1196 Gland
Switzerland 002-649-11

Living Oceans

Living Oceans is a division of the National Audubon Society dedicated to encouraging people to work to restore abundant marine wildlife and healthy habitats in our oceans and along our coasts. More than eighty Audubon chapters and over five hundred individuals have joined this activist group whose long-range goal is to develop local and regional committees of oceans activists who will identify crucial marine and coastal conservation issues in their areas and then work with land-use planners and community leaders to address them. Living Oceans is supported by grants from the Pew Charitable Trusts and the Packard, Munson, Kendal, and Norcross Foundations. Living Oceans publishes quarterly newsletters, *Audubon Activist,* and *Living Oceans News.*
550 South Bay Avenue
Islip, New York 11751
(516) 859-3032

The Marine Conservation Network

A volunteer-based organization which works with local scuba divers along the central and north coasts of California in underwater monitoring activities, including programs that will support and complement current efforts within the Monterey Bay National Marine Sanctuary. MCN provides volunteer divers with specialized training, standardized log sheets, and a regular publication. In addition to marine surveys, MCN sponsors informational lectures concerning current marine ecology issues and is working to establish hands-on marine ecology programs for children, including snorkeling, tide pool exploration, and cleanups.
PO Box 1854
Danville, CA 94526
(510) 838-2544

Marine Fish Conservation Network

A non-profit coalition of fishing and conservation organizations representing more than five million U.S. citizens. Disseminates educational material on marine resource and habitat conservation. Conducts lobbying workshops and letter-writing campaigns on issues such as overfishing, fisheries management, bycatch regulations, and long-term sustainability of marine resources. Current emphasis is on amending the Magnuson Fishery Conservation and Management Act.
1725 DeSales St. NW
Washington, DC 20036
(202) 857-3274

Marine Spill Response Corporation
A non-profit organization which cleans up oil spills around the U.S.; also publishes a monthly newsletter, *Response Log*, and staffs an Office of Public Affairs and divisions for education, research, and development.
1350 I St., NW, Suite 300
Washington, DC 20005
(202) 408-5700

National Association of Conservation Districts
This organization advances the interests of the U.S. conservation districts and provides needed services to further the conservation, management, and responsible development of natural resources.
PO Box 855
League City, TX 77574-0855
(800) 825-5547

National Audubon Society
Research, education, and lobbying on wildlife, forests, wilderness, public lands, endangered species, water, and energy policy. Publishes bimonthly magazine *Audubon* and *Audubon Activist* newsletter.
950 Third Avenue
New York, NY 10022
(212) 832-3200

National Coalition for Marine Conservation (NCMC)
Founded in 1973, the NCMC is an independent, non-profit organization of fishermen and environmentalists which uses education and activism to protect the marine environment. NCMC's goals are to: 1) restore depleted fish populations to healthy levels; 2) promote sustainable use policies that balance commercial, recreational, and ecological values; 3) eliminate wasteful fishing practices; 4) improve our understanding of fish and their role in the marine ecosystem; and 5) preserve coastal habitat and water quality.
5105 Paulsen Street #243
Savannah, GA 31405
(912) 354-0441

National Toxics Campaign
Seeks to prevent pollution, protect public health, provide guidance for communities, and laboratory testing for air and water pollution, and pressures corporations to clean up and manufacture safer products.

1168 Commonwealth Ave.
Boston, MA 02134
(617) 232-0327

National Wildlife Federation
With a network of fifty-one state and territorial affiliates, NWF pro-
motes the wise use of natural resources. Sponsors National Wildlife
Week and many other educational and demonstration programs.
Publishes *National Wildlife* bimonthly, *Nature Scope*, *Environmental
Quality Index,* and *Legislative Hotline.*
1400 16th Street, NW
Washington, DC 20036-2266
(202) 797-6800

Natural Resources Defense Council
A national organization dedicated to protecting the natural and
human environment. It combines research, education, advocacy,
and litigation on toxic substances, air and water pollution, nuclear
safety, and other issues. Publishes a bimonthly newsletter and the
quarterly *Amicus Journal.*
1015 31 Street, NW
Washington, DC 20007
(202) 333-8495

The Nature Conservancy
An international organization dedicated to preserving plants, animals, and natural communities that represent the diversity of life on
earth by protecting their natural habitats. The Nature Conservancy
manages a system of over 1,300 nature sanctuaries in the fifty states,
and assists non-governmental organizations abroad with environmental initiatives.
1815 North Lynn Street
Arlington, VA 22209
(703) 841-5300

Ocean Alliance
Established in 1978, Ocean Alliance's goal is to protect water sources
and their plant and animal life. The Alliance promotes conservation,
education, and research and helped to establish the National Marine
Sanctuaries in California, the Gulf of the Farallones, Cordell Bank, and
the Channel Islands.
Building E, Fort Mason Center
San Francisco, CA 94123
(415) 441-5970

Pacific Whale Foundation

The Pacific Whale Foundation was established in 1980 to conduct scientific research and educational programs and to promote public awareness of whales, dolphins, and porpoises. The organization's 5,000 members include conservationists, scientists, and volunteers.
Kealia Beach Plaza
101 N. Kihei Road, Suite 21
Kihei, Maui, HI 96753-2615
(808) 879-8860

Population-Environment Balance

A non-profit activist organization dedicated to creating a sustainable environment through controlling U.S. population. Currently this group of 5,000 members is active in lobbying and supporting legislation that reduces illegal and legal immigration and replacement level family planning. Educational materials are available to members.
1325 G Street NW, Suite 1003
Washington, DC 20005-3104
(202) 879-3000

PADI Project AWARE Foundation

A leading foundation in the diving industry which funds and supports environmental research, education, and protection. Publishes the quarterly *Undersea Journal*.
1251 East Dyer Rd., Suite #100
Santa Ana, CA 92705-5605
(714) 540-7234

Project Reefkeeper

A non-profit organization focusing on coral reef conservation. Specific issues include offshore oil, marine pollution, living marine resources, physical impacts, and protected areas. Project Reefkeeper works to effect legislation, conducts educational workshops, and publishes *Reefkeeper Report* bi-monthly and *Reef Alert* several times a year. The Reefkeeper Network, whose worldwide membership is 9,000, interfaces with and offers technical support to other environmental groups.
1800 SW 1st St., Suite 306
Miami, FL 33135
(305) 642-9443

Reef Relief
A non-profit organization working to preserve and protect the coral reefs of the Florida Keys. Reef Relief has developed a public education and outreach program for protecting coral reefs and maintaining mooring buoys.
PO Box 430
Key West, FL 33041
(305) 294-3100

Renew America
Renew America was established as a clearinghouse for environmental information. The organization disseminates information and gives recommendations to policy makers, environmental organizations, and the media. One of Renew America's biggest projects is the annual State of the States report.
1400 Sixteenth Street, NW, Suite 710
Washington, DC 20036
(202) 232-2252

Save the Whales, Inc.
An organization which educates children and adults about marine mammals, their environment, and their preservation. "Whales on Wheels" program provides educational materials and programs via a mobile unit. Save the Whales offers lectures, publishes a quarterly newsletter, promotes letter-writing campaigns, and media appearances, and supports marine mammal research in the wild.
PO Box 2397
Venice, CA 90291
(310) 392-6226

Sea Shepherd Conservation Society
An outspoken direct-action group which seeks to protect marine animals and marine habitats. Some of their highly publicized efforts have been focused on preventing the killing of dolphins by the tuna industry, enforcing a moratorium on whaling, and the rescue of whales and marine mammals in international waters.
PO Box 7000-S
Redondo Beach, CA 90277
(213) 373-6979

Sierra Club
The nation's oldest and still very effective voice for the preservation of the ecosystems and resources of the earth. Its work includes lob-

bying, public education, political action, grass-roots organizing, and national and international group outings and tours. Publishes the bimonthly magazine *Sierra*.
730 Polk Street
San Francisco, CA 94109
(415) 776-2211

Soil and Water Conservation Society
An organization which advocates the conservation of soil, water, and related natural resources.
7515 NE Ankeny Rd.
Ankeny, IA 50021-9764
(515) 289-2331

Surfrider Foundation
Founded in 1984, a non-profit coastal activist membership organization with twenty chapters in California, Hawaii, New York, New Jersey, Virginia, North Carolina, Florida, and Australia. The goals are to protect and enhance the world's waves and beaches through conservation, research, and education.
PO Box 2704, 86
Huntington Beach, CA 92647
(714) 960-8390 (800) 743-SURF

United Nations Environment Programme
The environmental agency of the United Nations was created in 1972. Its world headquarters is in Nairobi, Kenya, but the staff of over two hundred people are spread throughout the world. This organization helps formulate and coordinate environmental policies at the municipal, national, regional, and international levels by working with scientific agencies, the private and public sectors, non-governmental organizations, and legal institutions.
2 United Nations Plaza, Room DC2-303
New York, NY 10017
(212) 963-8093

Water Environment Federation
A non-profit international educational and technical organization of 40,000 water quality experts. Specialists include civil, chemical, and environmental engineers, biologists, government officials, treatment plant managers and operators, lab technicians, college professors, and equipment manufactures and distributors. The Water Environment Federation's mission is the preservation and enhancement of water quality worldwide. WEF publishes numerous jour-

nals, newsletters, and manuals of practice.
601 Wythe Street
Alexandria, VA 22314-1994
(703) 684-2400

The Wilderness Society

With thirteen regional offices, this organization educates citizens, public officials, and media on the need to protect and carefully manage public lands. Testifies at congressional hearings. Sponsors meetings on public land management. Publishes wilderness-related reports.
900 17th St. NW
Washington, DC 20006
(202) 833-2300

Wildlife Conservation International

This organization helps preserve the earth's biological diversity and valuable ecosystems. With 140 projects in forty-five countries, WCI addresses conflicts between humans and wildlife and explores sustainable solutions.
c/o New York Zoological Society
(Bronx Zoo/Brooklyn Aquarium)
New York, NY 10460
(718) 220-6891

The Wildlife Society

The Wildlife Society is a non-profit scientific and educational organization dedicated to conserving and sustaining wildlife productivity and diversity through resource management and to enhancing the scientific and technical capacity and performance of wilderness professionals.
5410 Grosvenor Lane
Bethesda, MD 20814-2197
(301) 897-9770

World Wildlife Fund

This organization works worldwide to preserve endangered wildlife and wildlands by encouraging sustainable development and the preservation of biodiversity. WWF is affiliated with the international WWF network, which has representatives in 40 countries. WWF publishes a bi-monthly newsletter, *Focus*.
1250 24th ST. NW, Suite 400
Washington, DC 20037
(202) 293-4800

APPENDIX 4
RECOMMENDED READING AND
AV MATERIALS

BOOKS

Bulloch, David K. *The Wasted Ocean: The Ominous Crisis of Marine Pollution and How to Stop It.* New York: Lyons and Burford, 1989. *The Underwater Naturalist.* New York: Lyons & Burford, 1991.

Carson, Rachel. *The Edge of the Sea.* Boston: Houghton-Mifflin, 1955.

Corson, Walter. *The Global Ecology Handbook: What You Can Do about the Environmental Crisis.* Boston: Beacon Press, 1990.

Cousteau, Jacques-Yves. *The Cousteau Almanac: An Inventory of Life on Our Water Planet.* New York: Doubleday, 1981. *The Ocean World.* New York: Harry N. Abrams, Inc., 1985.

Frank, Irene, and David Brownstone. *The Green Encyclopedia.* New York: Prentice Hall, 1992.

Gore, Albert. *Earth In The Balance: Ecology and the Human Spirit.* Boston: Houghton Mifflin, 1992.

Government Institutes, Inc. *Environmental Telephone Directory, 1992-1993.* Rockville, MD.

Lovelock, J.E. *Gaia: A New Look at Life on Earth.* Oxford University Press, 1975.

Maraniss, Linda. *All About Beach Cleanups.* Washington, DC: Center for Marine Conservation, 1989.

Marx, Wesley. *The Frail Ocean.* Ballantine Books, 1970.

Meadows, Donnella, Dennis Meadows, and Jorgen Randers. *Beyond the Limits: Confronting Global Collapse, Envisioning a Sustainable Future.* Post Mills, VT: Chelsea Green, 1992.

Meyer, Christine, and Faith Moosang. *Living with the Land: Communities Restoring the Earth.* Philadelphia: New Society Publications, 1992.

Millemann, Beth. *And Two If By Sea: Fighting the Attack on America's Coasts.* Washington, DC: Coast Alliance, 1987.

Naar, John.   *Design for a Livable Planet.* New York: Harper
                       and Row, 1990.
O'Hara, Kathy.   *A Citizen's Guide to Plastics in the Ocean: More
                       Than a Litter Problem.* Washington DC: Center for
                       Marine Conservation, 1988.
Thorne-Miller,   *The Living Ocean: Understanding and Protecting
                       Marine*
Boyce, and.   *Biodiversity.* Washington DC: Island Press, 1991.
John Catena
Safina, Carl.   *A Primer on Conserving Marine Resources.* Islip,
                       New York: National Audubon Society, 1994.
Stein, Edith.   *The Environmental Sourcebook.* New York: Lyons
and Burford,   1992.
U.S. Environmental   *Marine and Estuarine Protection: Programs and
Protection Agency.   Activities.* 1989.
Weber, Michael *et al.* *The 1985 Citizen's Guide to the Ocean.* 1986.

ARTICLES AND PAPERS

Brown, Lester *et al.* *State of the World 1992,* New York: W.W.Norton,
                       1992.
Curtis, C.   "Protecting The Oceans." *Oceanus* 33 (2): 19-28.
                       "Don't Go Near the Water." *Newsweek,* 1 August
                       1988, 42-48.
Kitsos, T.,   "Congress and Waste Disposal at Sea." *Ocean-us*
and J. Bondareff.   33 (2): 23-28.
Laycock, George.   "Good Times are Killing the Keys." *Audubon*
                       (Sept./Oct. 1991): 38-49.
Lessen, Nicholas.   "The Ocean Blues." *World Watch* 2 (4).
"Our Filthy Seas."   *Time,* August 1988, 44-50.
National Academy   "Marine Litter." In *Assessing Potential Ocean
of Sciences.   Pollutants.* Washington, DC: National Research
                       Council, 1974, 405-438.
Spenser, D.   "The Ocean and Waste Management." *Oceanus*
                       33 (2):   5-11.
Talge, Helen.   "Impact of Recreational Divers on Coral Reefs in
                       the Florida Keys." *Diving for Science-AAUS
                       Proceedings 1990:* 356-373.
                       "The State of the Marine Environment." *United
                       Nations Environmental Program News*
                       (April 1988).
Viders, Hillary.   "Where's the Reef?" *Sources* (Nov./Dec. 1990):
                       22-28.

Ward, Fred.   "Florida's Coral Reefs are Imperiled." *National*
         Geographic (July 1990): 115-132.
Weber, Peter.   *Abandoned Seas: Reversing the Decline of the*
         Oceans. Washington, DC: Worldwatch Institute.
Weiskopf, Michael. "Plastic Reaps a Grim Harvest in the Oceans of
         the World." *Smithsonian* (March, 1988): 58-66.

FILMS AND VIDEOS

Caribbean Reef Encounters, Delphin Productions, 1994
Discover the Underwater World, PADI
The Living Ocean, National Geographic Films, 1988
Reef Dive, NOAA/National Park Service, 1988
The Florida Sanctuaries, NOAA/National Park Service, 1988
Inherit the Sea, Center for Marine Conservation, 1991
Peak Performance Buoyancy, PADI
Piggy Divers, YMCA National Scuba Program
To Preserve and Protect, NAUI/NOAA
You Can Make A Difference, NAUI/NOAA

NEWSLETTERS AND PERIODICALS

Audubon Activist, National Audubon Society
The Calypso Log, The Cousteau Society
Coastal Connection, Center for Marine Conservation
Heritage, Mutual of Omaha
Living Oceans News, National Audubon Society
Marine Conservation News, Center for Marine Conservation
National Geographic, National Geographic Society
Oceanus, Woods Hole Oceanographic Institution
Ocean Watch, Oceanic Society
PADI Undersea Journal, PADI
Reef Keeper Alert, Project Reefkeeper
Sanctuary Currents, Center for Marine Conservation/NOAA
Sources, National Association of Underwater Instructors
The Underwater Naturalist, Project Reefkeeper, American Littoral Society

APPENDIX 5

ORGANIZATIONS AND AQUARIUMS TO CONTACT FOR ENVIRONMENTAL EDUCATIONAL MATERIALS

The following organizations provide information about marine science and education. Some have materials specifically for school curricula.

Alliance for Environmental Education
PO Box 368
The Plains, VA 22171
(703) 253-5812

American Cetacean Society
PO Box 2639
San Pedro, CA 90731
(213) 548-6279

American Littoral Society
Sandy Hook
Highlands, NJ 07732
(201) 291-0055

American Society for Environmental Education
58 Main Street
Box R
Durham, NH 03824
(603) 868-5700

Caribbean Conservation Corporation
P.O. Box 2866
Gainsville, FL 32602
(904) 373-6441 (800) 678-7853

Center for Environmental Information
46 Prince Street
Rochester, NY 14607
(716) 271-3550

Center for Marine Conservation
1725 DeSales St. NW
Washington, DC 20036
(202) 429-5069

Chesapeake Bay Foundation
162 Prince George St.
Annapolis, MD 21401
(301) 268-8816

Cousteau Society
930 West 21st St.
Norfolk, VA 23517
(804) 627-1144

The Environmental Exchange
1930 18th St. NW, Suite 24
Washington, DC 20009
(202) 387-2182

Friends of the Sea Otter
PO Box 221220
Carmel, CA 93922
(408) 625-3290

Inform
381 Park Ave. S.
New York, NY 10016
(212) 689-4040

International Oceanographic Foundation
3979 Rickenbacker Causeway, Virginia Key
Miami, FL 33149-9900
(305) 361-5786

Marine Fish Conservation Network
1725 DeSales St. NW
Washington, DC 20036
(202) 857-3274

Monterey Aquarium
886 Cannery Row
Monterey, CA 93940
(408) 648-4888

National Aquarium Baltimore
Pier 3/501 E. Pratt ST.
Baltimore, MD 21202
(301) 727-3000

National Association of Biology Teachers
11250 Roger Bacon Drive #19
Reston, VA
(703) 471-1134

National Marine Educators Association
PO Box 51212
Pacific Grove, CA 93950

National Science Teachers Association
1742 Connecticut Ave, NW
Washington, DC 20009-1171
(202) 328- 5800

National Wildlife Federation
1400 16th Street, NW
Washington, DC 20036-2266
(202) 797-6800

New England Aquarium
Teacher Resource Center
Central Wharf
Boston, MA 02110-3309
(617) 973-5200

New Jersey State Aquarium
1 Riverside Drive
Camden, New Jersey 08103
(609) 365-3300

NOAA
Sea Grant Program R-OR 1
1335 East/West Highway
Silver Springs, MD 20910-3226
(301) 713-2431

North American Association for Environmental Education
PO Box 400
Troy, Ohio 45373

PADI Project AWARE
1252 East Dyer Rd., Suite 100
Santa Ana, CA 92705-5605
(714) 540-7234

Reef Relief
PO Box 430
Key West, FL 33041
(305) 294-3100

Save the Whales, Inc.
PO Box 2397
Venice, CA 90291
(310) 392-6226

Sea Education Association
PO Box 6
Church St.
Woods Hole, MA 02543
(617) 540-3954

Sierra Club
730 Polk Street
San Francisco, CA 94109
(415) 776-2211

APPENDIX 6

DIVE TRAINING AGENCIES

**International Diving
Educators Association (IDEA)**
PO Box 17373
Jacksonville, FL 32245
(904) 744-5554

International Scuba Educators
PO Box 17388
Clearwater, FL 34622-0388
(813) 539-6491

**Los Angeles County Dept. of
Parks and Recreation**
Underwater Unit
419 East 192 Street
Carson, CA 90746
(310) 217-8376

**National Association of
Underwater Instructors (NAUI)**
4650 Arrow Highway, Suite F-1
Montclair, CA 91763-1150
(909) 621-5801

**National Association of Scuba
Diving Schools (NASDS)**
1012 S. Yates
Memphis, TN 38119
(901) 767-7265

**Professional Association
of Diving Instructors (PADI)**
1251 East Dyer Rd. Suite 100
Santa Ana, CA 92705-5605
(714) 540-7234

Scuba Schools Int. (SSI)
2619 Canton Court
Ft. Collins, CO 80525
(303) 482-0883

**National Association of
Scuba Educators (NASE)**
1728 Kingsley Ave. #6
Orange Park, FL 32073
(904) 264-4104

**YMCA National Scuba
Program**
6083-A Oakbrook Pkwy
Norcross, GA 30093
(404) 242-9059

FOOTNOTES

Chapter 3

[1]"Fantastic World of Creatures is Found in Sea's Middle Depths." *New York Times,* July 26, 1994.

Chapter 7

[1]Foster and Schiel, *The Ecology of Giant Kelp Forests in California.*

Chapter 10

[1]Foster *et al.* 1990. *Northwest Environmental Journal* 6: 105-120.

[2]Miller, G. *Living in the Environment.* 6th ed., (Belmont, CA: Wadsworth Publishing Co., 1992), 12.

[3]Ibid., 13.

Chapter 13

[1]*Skin Diver Magazine 1994 Survey* (Los Angeles, CA: Peterson Publishing).

[2]Although there is presently no cooperative coalition among these various sectors, the author and several key industry officials are in the process of developing one. In conjunction with this coalition will be an industry-wide environmental directory to facilitate the collaboration of resources and funding for conservation projects.

Chapter 18

[1]"Can Ecotourism Save the Planet?" *Conde Nast Traveler* (Dec. 1994): 35-143.

Appendix 1

[1]In 1972, after several cases of leukemia and deformities were traced to dioxin contamination in the ground water, the entire community of Love Canal (near Niagara Falls, NY) had to be evacuated. It was revealed that the property had been used in the 1940s as a toxic waste dump. The enormous task of cleaning up Love Canal has cost the government many billions of dollars and is still in progress.

INDEX

Q

R

S

Y

Z